I0493839

<u>Disclaimer</u>

Book Title: Performance Analysis of RF-Based Electronic Safety Equipment in a Subway Station and the Empire State Building

Book Author: William F. Young; Catherine A. Remley; Galen H. Koepke; Dennis G. Camell; Jacob L. Healy;

Book Abstract: We analyze data from NIST field tests in which radio-propagation channel path loss values were measured at approximately the same physical locations where the performance of various RF-based firefighter distress beacons were tested. These side-by-side tests were made in two key representative emergency responder environments, a New York subway station and the Empire State Building. These environments contain propagation features that may impair radio communications, including stairwells, tunnels, and rooms deep within buildings, among others. The goal of this work is to determine appropriate performance metrics for use in the development of laboratory-based test methods for RF-based electronic safety equipment. The analysis supports the classification of structures into categories of attenuation values that can be used in laboratory-based test methods to verify the performance of the RF-based alarm systems. The environments, tests, and measured data are discussed in detail. The RF propagation-channel data also provide insight into the expected attenuation in high-rise buildings and below-ground structures.

Citation: NIST TN - 1792

Keywords: attenuation; emergency responders; firefighter communications; public-safety radio communications; radio propagation-channel measurements; wireless communications

NIST Technical Note 1792

Performance Analysis of RF-Based Electronic Safety Equipment in a Subway Station and the Empire State Building

William F. Young
Kate A. Remley
Galen Koepke
Dennis Camell
Jacob Healy

http://dx.doi.org/10.6028/NIST.TN.1792

National Institute of
Standards and Technology
U.S. Department of Commerce

NIST Technical Note 1792

Performance Analysis of RF-Based Electronic Safety Equipment in a Subway Station and the Empire State Building

William F. Young
Kate A. Remley
Galen Koepke
Dennis Camell
Jacob Healy
Electromagnetics Division
Physical Measurement Laboratory

http://dx.doi.org/10.6028/NIST.TN.1792

March 2013

U.S. Department of Commerce
Rebecca Blank, Acting Secretary

National Institute of Standards and Technology
Patrick D. Gallagher, Under Secretary of Commerce for Standards and Technology and Director

National Institute of Standards and Technology Technical Note 1792
Natl. Inst. Stand. Technol. Tech. Note 1792, 70 pages (March, 2013)
http://dx.doi.org/10.6028/NIST.TN.1792
Article I.   CODEN: NTNOEF

Contents

Executive Summary

The National Institute of Standards and Technology (NIST) has participated in a multi-year project to support the development of performance metrics and test methods for radio-frequency (RF)-based electronic safety equipment used by the public-safety community. The work reported here focuses on side-by-side measurements of radio-propagation environment characteristics and actual wireless device performance in two key representative emergency responder environments, a subway station and a high-rise building. Identifying the impacts of path loss on wireless device performance in various environments enables the development of standardized laboratory-based test methods that simulate the conditions under which electronic safety equipment will be used in the field. The test methods can then be incorporated into consensus standards for this equipment.

The analysis presented here has been funded by the U.S. Department of Homeland Security Standards Branch. The work reported here focuses on RF-based personal alert safety systems (PASS), used by firefighters to indicate when a firefighter is motionless or in distress. However, the methodology, analysis, and RF propagation results presented here could easily be extended to other types of wireless devices that operate in similar environments.

In previous propagation-channel studies, NIST engineers measured path loss ("attenuation") and the level of reflectivity (or "multipath," quantified by the root-mean-square delay spread) in large public structures and environments where radio communications could be challenging. (See NIST Technical Notes 1545, 1456, 1550, 1552, 1557, and 1559.) These environments included multi-story buildings; buildings with subterranean floors and tunnels; buildings with deep interior spaces; those with few windows; and outdoor "urban canyons," consisting of city streets surrounded by tall buildings. The NIST Public-Safety Communications Research Lab funded those measurements of the propagation channel. Current work focuses on the subterranean environment presented by subways, and a 100-story high-rise, the Empire State Building. The tests were completed in New York City with the help of the Fire Department of New York and several firefighters from the Research and Development Division. Access to the subway station was provided by the New York City Transit Authority. The building owners, management, and maintenance personnel provided access and support for testing the Empire State Building.

To support standards development in public-safety applications, the NIST studies focus on the penetration of radio signals from outside to inside a given structure (and vice versa), as opposed to outdoor-to-outdoor or within-building tests. To simulate an incident command post in the propagation-channel studies, a receive antenna was positioned outside of each structure at a location representative of a fireground configuration. Portable RF PASS units were carried through the environment and success or failure in receipt of various alarms was noted. A continuous-wave radio transmitter was then carried through the structures with a spectrum-analyzer-based receiver system continually measuring the received signal power. Key locations in the walked path were "marked" in the collected data. These key location "markers" enable the correlation between path loss and RF PASS performance at specific locations within the structure. The markers also allow parsing of the data among different floors or platform levels, which supports path-loss analysis on a per-floor or level basis.

The RF-based PASS systems are capable of two-way communications, and these measurements determined whether or not the alarm from the portable RF PASS device was received by the base station and if the portable device received an alarm signal from the base station. The portable RF PASS device generally transmits with a lower power than the base station in order to conserve battery life and, in some cases, to meet "intrinsic safety" standards for electronic equipment. Consequently, the signal emitted from the RF-based PASS device is typically weaker, and testing the reception of this alarm by the base station represented a worst-case scenario.

A second aspect of these tests was the collection of path-loss data for these two high-attenuation RF propagation-environments at 430 MHz, 750 MHz, 905 MHz, 1834 MHz, and 2405 MHz. With the simulated incident command post located outside, but near an entrance to the structure, the maximum measured path-loss values ranged from 140 dB to 175 dB for the Empire State Building and 210 dB to 240 dB for the subway station. The 140 dB path-loss value suggests that a point-to-point system could provide coverage over much, though not necessarily all, of this RF propagation-environment. The range of success with one of the point-to-point systems during the RF PASS systems testing verified that good, but not complete, coverage was indeed possible. If an RF PASS system is deployed with the incident command post outside a large structure, even without subterranean elements, the encountered path-loss should be expected to be at least 140 dB to 175 dB. A large, subterranean structure will likely exhibit path-loss values greater than 200 dB.

We tested four different commercially available RF-based PASS systems, one that operates on a licensed frequency in the 450 MHz public-safety narrowband frequency allocation, two that operate in the unlicensed spectrum between 902 MHz and 928 MHz, and one that operates in the unlicensed spectrum of 2.4 GHz. In the case of the subway environment, one 900 MHz and the 2.4 GHz systems were tested with and without a repeater unit. We expect that use of repeater technology will continue to increase in RF-based systems. The tests demonstrated that the repeater approach added significant improvement coverage for those two systems.

The data and corresponding discussion presented below are intended to aid in the development of laboratory-based test methods for RF-based emergency safety equipment such as RF-based PASS devices. Test methods developed to date focus on inserting a controllable amount of attenuation (i.e., 100 dB of path-loss) between the portable PASS device and the PASS base station and inserting a specified level of RF interference between the portable and base station units. These current tests have been adopted in the 2013 revision of NFPA 1982: Personal Alert Safety Systems. We anticipate that additional test methods and standards covering multi-hop networks and medium-to-high attenuation environments will be forthcoming in the near future as well. The data and test results provided here directly support the development of these future tests and standards.

Performance Analysis of RF-Based Electronic Safety Equipment in a Subway Station and the Empire State Building

William F. Young, Kate A. Remley, Galen Koepke, Dennis Camell, and Jacob Healy

Electromagnetics Division
National Institute of Standards and Technology
325 Broadway, Boulder, CO 80305

Abstract: *We analyze data from NIST field tests in which radio-propagation channel path-loss values were measured at approximately the same physical locations where the performance of various RF-based firefighter distress beacons were tested. These side-by-side tests were made in two key representative emergency responder environments: a New York subway station and the Empire State Building. These environments contain propagation features that may impair radio communications, including stairwells, tunnels, and rooms deep within buildings. The goal of this work is to determine appropriate performance metrics for use in the development of laboratory-based test methods for RF-based electronic safety equipment. The analysis supports the classification of building structures into categories of attenuation values that can be used in laboratory-based test methods to verify the performance of the RF-based alarm systems. The environments, tests, and measured data are discussed in detail. The RF propagation-channel data also provide insight into the expected attenuation in high-rise buildings and below-ground structures.*

Key words: *attenuation; emergency responders; firefighter communications; public-safety radio communications; radio propagation-channel measurements; wireless communications.*

1. Introduction

Emergency responders count on reliable radio communications between responders, who are often inside a structure, and the incident command station outside. New wireless technology is being developed that can further increase responders' safety and efficiency by remotely monitoring their position, status, and situational awareness. The responder community would like to take advantage of this technology. Because lives may depend on its performance, wireless technology used in emergency response scenarios must generally satisfy higher levels of reliability than technology used in the commercial sector. Even though standards currently exist for commercial wireless devices such as cell phones, wireless local-area-networks and handheld radios, few standards currently exist for wireless electronic safety equipment. Unlike commercial-sector applications, the performance requirements for electronic safety

1

equipment typically focus on communication reliability and availability rather than the amount of data throughput, and thus any applicable standards should reflect these key performance metrics.

The U.S. Department of Homeland Security (DHS) Standards Branch has tasked researchers at the National Institute of Standards and Technology (NIST) with providing technical support for the development of consensus standards for these new products. As examples, DHS, through NIST, has determined gaps in existing standards and developed appropriate test methods for RF identification (RFID) systems used in public-safety and government applications such as tracking or inventory control [1-3]. A second project is working with the urban search and rescue community to support development of standards for the wireless control of robots through ASTM International [4], [5].

Here we describe DHS-sponsored work carried out to support the National Fire Protection Association (NFPA) in the revision of NFPA 1982: Standard on Personal Alert Safety Systems (PASS) [6] to include RF-based PASS (RF PASS). A PASS is essentially a "firefighter-down" alarm that emits a loud audible alarm when the wearer is motionless for 30 seconds. Some PASS manufacturers are now including an RF transceiver in the portable, body-worn PASS device to alert the incident command station. The transceiver is also capable of receiving a signal from the incident command station to evacuate. The work presented here is expected to apply to other types of RF-based electronic safety equipment as these become available.

The technical approach taken here involves two main measurement activities. First, we collect RF propagation-channel data by walking over a predetermined path in a structure with a transmitter emitting a continuous-wave (CW) signal, as done in our previous RF propagation-channel studies [7-9]   A spectrum analyzer at a location that simulates an incident command station measures the received-signal strength. The measured values are recorded, and subsequently processed, to estimate the path at distinct locations, such as the location of a portable RF PASS, and general sections of the structure, such as different floors.

In the second part of the experiment, several RF PASS technologies are tested at specific locations along the same path covered in the CW transmitter walk. The base stations of these PASS systems are co-located with the spectrum analyzer measurement equipment; i.e., the simulated incident command station. At the specific test locations on the path walked, the RF PASS devices are tested for their ability to successfully send a firefighter-down alarm to the base station, and to successfully receive an evacuation alarm from the base station. In post-processing, the RF propagation-channel data provide path-loss information that is correlated with the RF PASS test locations. This analysis supports the refinement of laboratory tests for RF PASS devices as well as RF-based emergency equipment, in general.

While the aforementioned testing offers insight into real-world RF PASS performance, there are limitations and challenges with the direct use of such environments as test beds. Laboratory-based test methods of RF-based emergency safety equipment provide the advantages of accuracy, repeatability, efficiency, and, often, reduced cost, when compared to the use of building structures and/or structure-based test beds. This is because it is possible to carefully control the test environment and conditions in a laboratory while covering propagation-channel parameters measured over a wide range of building types. For the testing of RF-based equipment, we can expose the system under test to specific and reproducible levels of attenuation, interference, or multipath, and with known uncertainties.

It has been necessary for NIST to perform the RF propagation measurements as part of this project because much of the data that have been previously published in the literature describe tests made to support commercial applications such as cellular telephone communications, where a cellular base station provides coverage to a wide area, rather than the point-to-point, pedestrian-height scenarios utilized in many RF-based emergency scenarios. In addition, one important goal of the study presented here was to analyze the performance of the RF PASS systems under the same conditions under which the channel characterization tests were conducted.

In the following sections, we discuss various aspects of our measurements and provide insights that support RF-based testing for similar technologies. We also discuss factors that contribute to the uncertainty in the measurement comparison. For instance, since the locations of the measurements were not identical, the performance of the RF PASS system does not, in some cases, agree with what is expected theoretically. However, certain trends are clearly indicated from the data allowing us to identify representative values of attenuation for the development of laboratory-based test methods.

In Section 2, we describe the various environments in which the measurements were performed, and include sketches of the two structures. Section 3 contains the RF PASS systems' performance results and analysis, including a discussion of the relationship between RF PASS performance and RF propagation-channel characteristics. We purposely provide RF PASS performance results before describing the RF propagation-channel measurements so that the reader is exposed to the system level performance first. This is because we anticipate most readers are interested in these system level performance results, with a smaller subset interested in the details of the RF propagation-channel itself. In Section 4, we describe the measurement system and data-processing algorithms used in the RF propagation-channel measurements. Section 5 describes the statistical methods used in representing the data, and Section 6 contains channel measurement results for the two environments. In Section 7, we discuss the assumptions and approximations that were made in measuring and analyzing the data and how they affect the uncertainty in relating RF-based PASS performance to propagation-channel characteristics. Finally, in Section 8, we provide conclusions based on these measurements and subsequent analysis.

2. Test Environments

We provide a brief description of the environments and conditions in which radio propagation-channel measurements were made and RF PASS systems were tested. Pictures of the general environment are included, along with basic diagrams of the subway station and Empire State building. Positions are marked on the diagrams of each structure where the tests of the RF PASS systems were conducted. Markers are also included for positions that delineate key sections of the structures such as floors, stairwells, and elevators. These sketches of the two structures are provided to give the reader a perspective on the relative locations of the RF PASS test points within the structures.

2.1 Subway station, New York City

The West 4th Street Station, with the entrance shown in Figure 1, was the subway station used for these RF PASS tests. An external receive site, Receive Site 1, was setup next to the entrance, with measurement equipment and the RF PASS base stations. Figures 2 and 3 show the Receive Site 1 setup along with some of the equipment used for the RF PASS tests. These pictures illustrate the close proximity of the site to the stairway entrance into the subway. The stairwell leading down to the pay turnstile level is shown in Figure 4. As shown in Figure 5, the stairwell is located down a narrow walkway from the pay turnstile. Inside the station, there are passenger platforms as shown in Figure 6.

Figure 1. The subway station where the RF PASS systems were tested.

Figure 2. FDNY and NIST personnel setting up Receive Site 1, located outside the subway entrance.

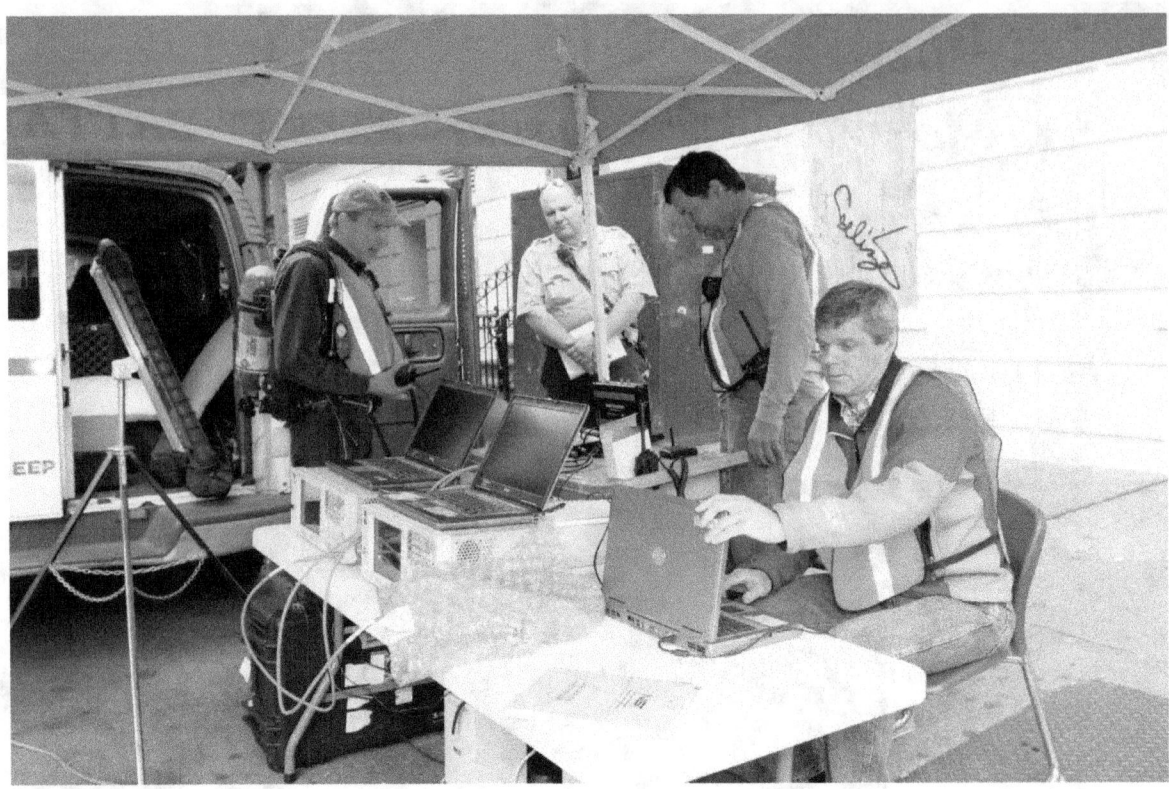

Figure 3. Receive Site 1 at the subway station. Some of the equipment in the picture includes spectrum analyzers and RF PASS base stations.

Figure 4. Subway station stairwell entrance.

Exit to street

Figure 5. Token booth turnstile. The stairwell that exits to the street is located at the end of the walkway.

Figure 6. Passenger platforms and stairwell between two of the platforms.

For the purposes of understanding the RF PASS tests and results, a general description of the subway layout is useful. The subway station consisted of three passenger platforms below ground, with a length of approximately three city blocks. A fare station located at an elevation between the street stairwell entrance and the first passenger platform connects the entrance via stairwells to the first platform. Thus, there are four distinct levels considered here, starting with the subterranean token booth (Level 1), the first subterranean passenger platform (Level 2), the middle, subterranean passenger platform or "Mezzanine Level" (Level 3), and the deepest subterranean passenger platform (Level 4). Figure 7 shows the basic layout, along with markers indicating the path walked with the radio used in the path-loss measurements. The order of the numbers indicates the progression in the walk. Points that are labeled with two numbers, such as point 7/19, indicate that the point was crossed both entering and exiting the subway station.

For RF PASS system testing, the base stations were located outside near the stairwell entrance at Receive Site 1. The second receive site, Receive Site 2, was used only for RF propagation-channel analysis. The RF PASS system results, mapped to the locations within the subway, are found in Section 3.

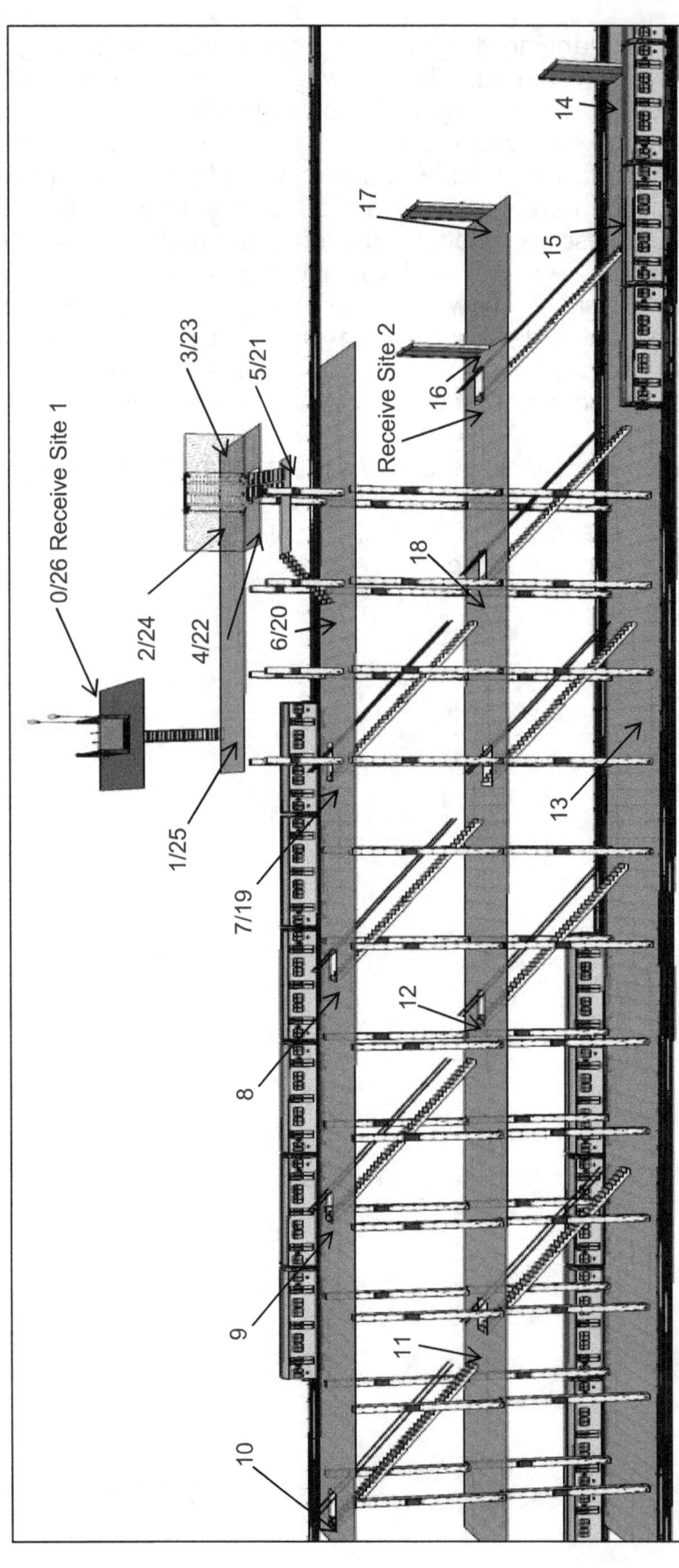

Figure 7. Layout of the four-level subway station. Receive Site 1 is located at the street level and Receive Site 2 is located on Level 3. The numbers indicate the marker locations corresponding to the propagation-channel measurement path. These locations are used to estimate the path loss at RF PASS test locations. Note that the figure is not to scale. RF PASS base stations were located at Receive Site 1.

10

2.2 Empire State Building, New York City

Tests were also conducted in the Empire State Building in New York. We provide a few pictures here of the building and receive site locations. Figure 8 shows FDNY and NIST personnel setting up the radio receive site and RF PASS base stations outside of the Empire State Building. In Figure 9, the main lobby and Receive Site 2 are shown. Figure 10 includes pictures from several floors with the building where the RF PASS tests were conducted. The floors were typically being remodeled or at least contained minimal furnishings.

Figure 11 shows a sketch of the Empire State Building with markers that indicate the progression of the radio walk-through used for path-loss measurements. As in the case of the subway, the walk begins on the sidewalk at Point 0, enters the building, and proceeds in a path that follows the increasing numbers. From Point 30 on the 83rd floor, the path returns to Point 0 in as direct of manner as possible, i.e., through a series of elevators and through the first floor. The transitions between different floors occurred largely via freight elevators; the entering and exiting locations were in the interior portion of the floors because the main elevator shaft is located in the core of the building.

The portable RF PASS devices were tested in specific locations where the RF propagation-channel was measured on several floors of the building. This supported the goal of obtaining correlated path-loss values and RF PASS performance. It also allowed comparison of path-loss behavior between floors in the building.

Figure 8. Setting up Receive Site 1 outside the Empire State Building.

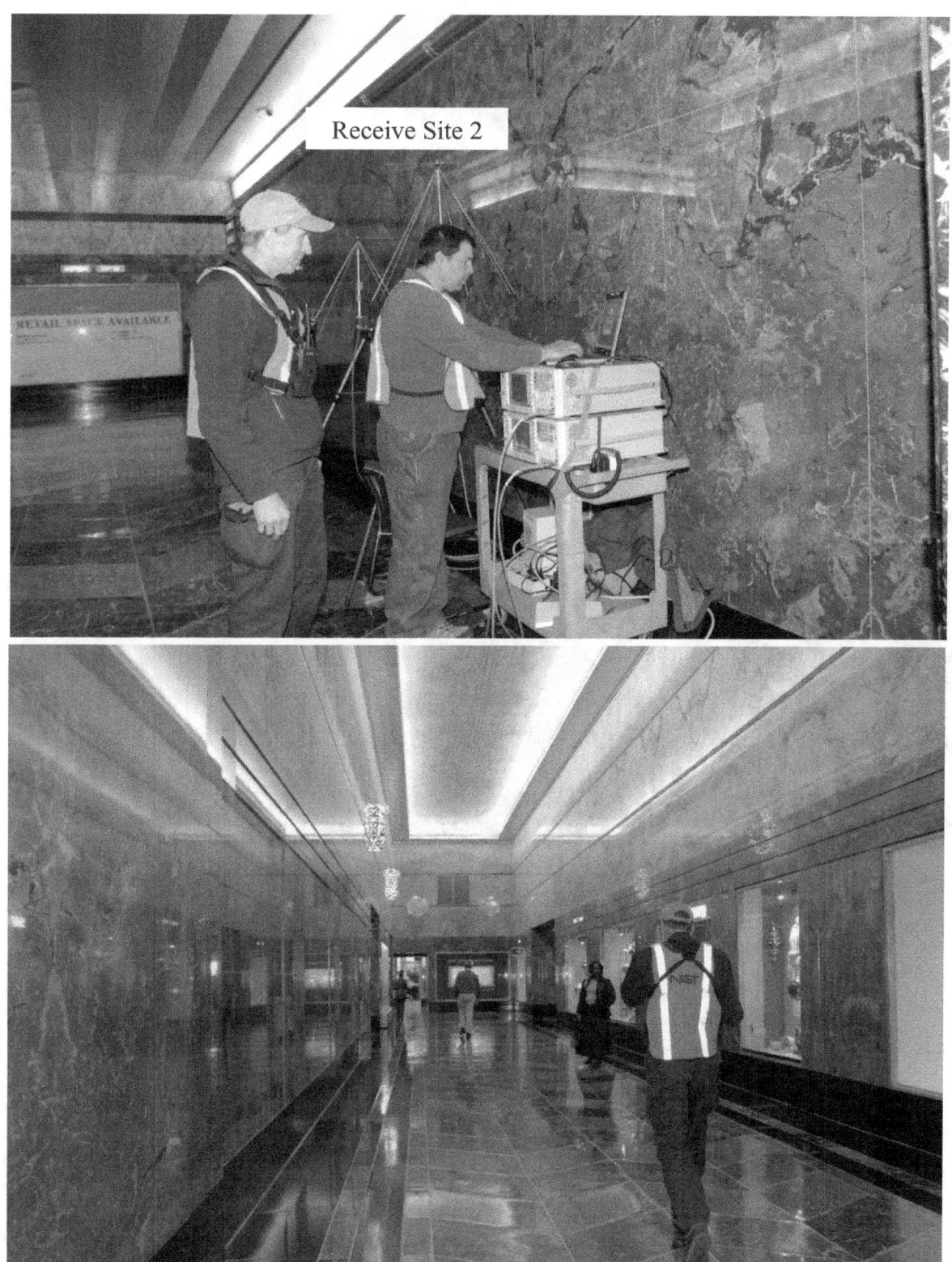

Receive Site 2

Figure 9. First floor lobby of the Empire State Building. Receive Site 2 was set up in the lobby.

13

Figure 10. Several of the floors and areas covered in RF PASS tests.

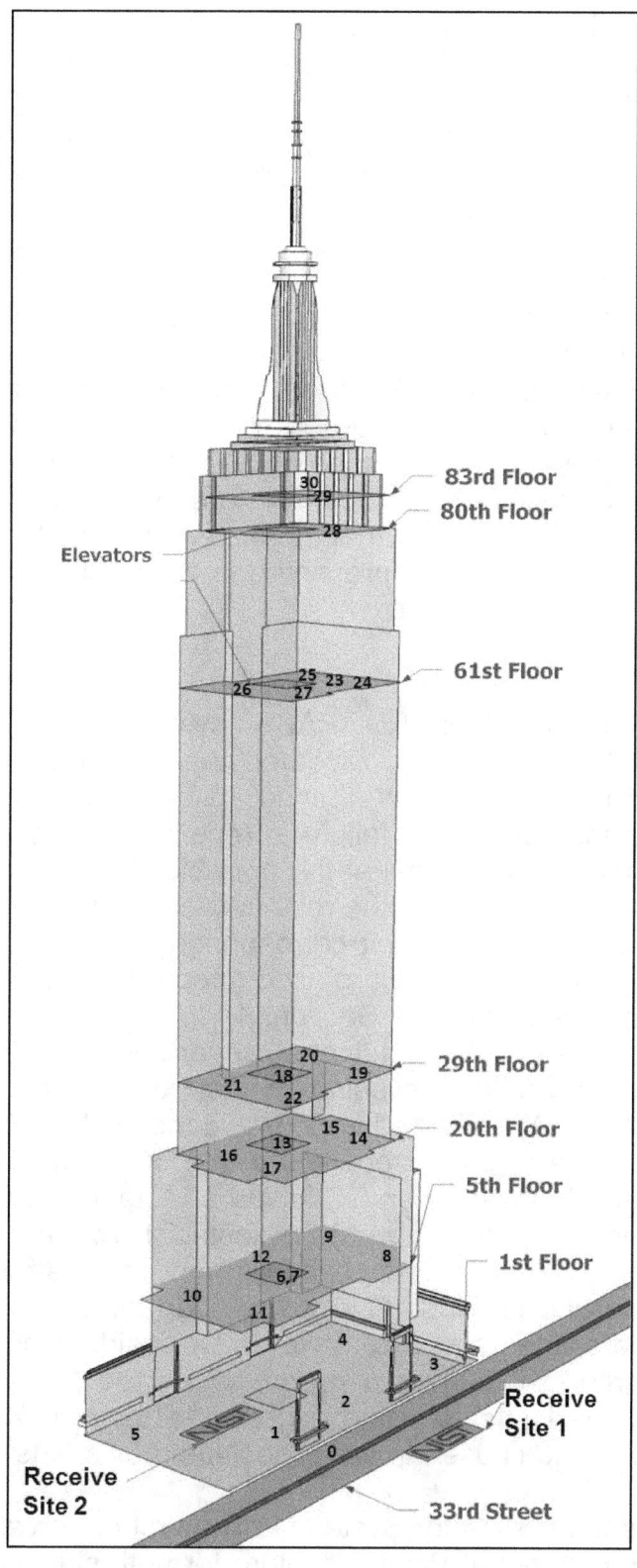

Figure 11. Empire State Building with receive/ base station sites depicted; Receive Site 1 is a parked van on the side of the road opposite the Empire State Building and Receive Site 2 is inside the building lobby. Marker numbers indicate the path walked with radio used for path-loss measurements. RF PASS base stations were located at Receive Site 1.

3. RF PASS System Performance in the Structures

This section describes results from RF PASS tests. First, we provide illustrations that show where a given RF PASS system either successfully or unsuccessfully executed a firefighter-down alarm. Second, we provide illustrations that show either the success or failure of RF PASS base-station initiated evacuation signals. Third, we include bar graphs that correlate the success/failure of the both alarm signals against the path loss in Section 4. For path-loss values at specific RF PASS test points in the subway station and Empire State Building, see Tables 7 and 8, respectively, in Appendix A.

The results in this section precede the RF propagation-channel results so that the reader is exposed to the system-level performance results before delving into the detailed analysis necessary to obtain RF propagation-channel values. We chose this order because we expect readers to range from those interested primarily in system-level performance, with minimal interest in the detail behind the RF propagation-channel measurements, to those interested in both system-level performance and detailed explanations of the RF propagation-channel measurement results. The former group of readers can concentrate on the results presented in this section, while the latter group will want to explore the entire document.

3.1 Subway station

Our results cover all four RF PASS systems tested for point-to-point communication. In addition, the built-in repeater capability of Systems 2 and 4 was tested with a single repeater. (Systems 1 and 3 did not have repeater capabilities). Figures 12 to 17 depict the success or failure in receiving a "firefighter-down" alarm at the base station or an evacuation signal at the portable PASS unit at specific locations in the subway station. The figures show the results overlaid on the sketch of the subway station, along with a marker number that corresponds to the RF propagation-channel test point location. The marker numbers are used to correlate the RF PASS results with path-loss measurements described in Section 4. If the system was tested with a repeater, two results at a test point are indicated, separated by a forward slash, "/".

These "firefighter down" and evacuation test results are labeled "Alarm" and "Evacuation," respectively. A measured success, denoted by a circle or "o" inside a rectangle, is indicated if the signal is received within 30 seconds; otherwise, the test was considered a failure, denoted by an "X" inside a rectangle. Due to the limited time available for field testing, some points were not tested. However, predicted results are indicated for some of the test points based upon the path-loss measurements, observations from these and previous field tests, and laboratory tests. A circle or "o" inside a hexagon is a predicted success, while an "X" inside a hexagon is a predicted failure. In addition, darker shading of a green "o" or red "X" indicates field-test or measured results, and the lighter shading a green "o" or red "X "indicates predicted results. This coloring scheme is used in the associated bar charts presented in Section 3.3.

Figure 12 shows the results for System 1. The first observation is that the Alarm and Evacuation signals both fail at Point 5. Failure for both signals is predicted starting at Point 6 and beyond because the measured path-loss is great than 125 dB. (For path-loss values at specific RF PASS test points in the subway station, see Table 7 in Appendix A.)

In Figures 13 and 14, results are shown for a point-to-point and a repeater configuration of System 2, respectively. Failure occurs starting at Point 2 for the point-to-point configuration, but the use of a repeater extends the range to Points 7, for the Alarm and Evacuation signals. At Point 8, both the Alarm and Evacuation signal fail. Clearly, System 2 experienced significant improvement in range when a repeater was used.

The System 3 results in Figure 15 indicate failure of the Alarm signal at Point 3 and beyond, and failure of the Evacuation signal at Point 4 and beyond. System 4 Alarm and Evacuation signal results in Figures 16 and 17, respectively, show a failure at Point 2 and beyond for both the Alarm and Evacuation signal if no repeater is used. When a repeater is placed at Point 1, both the Alarm and Evacuation signals are successful up to Point 6. Similar to System 2, the use of a repeater provides a significant improvement in range.

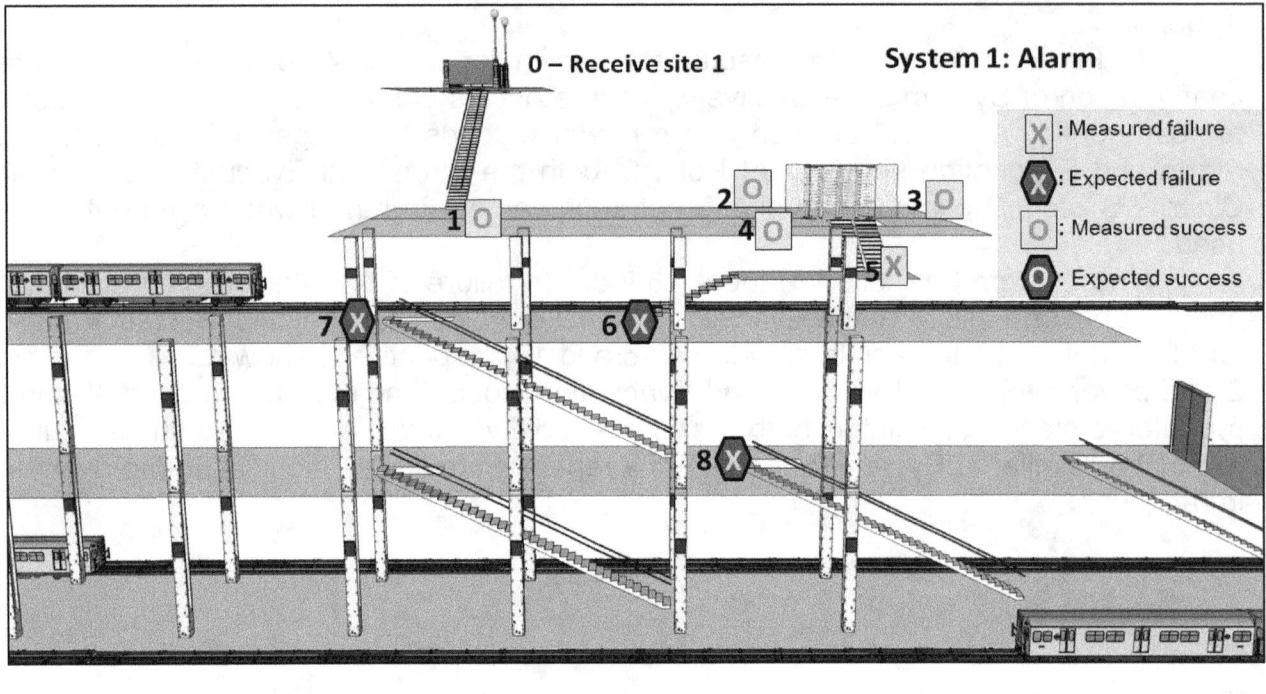

(a)

(b)

Figure 12. (a) Firefighter-down "Alarm" signal and (b) "Evacuation" signal for RF PASS System 1. System 1 did not have a repeater option.

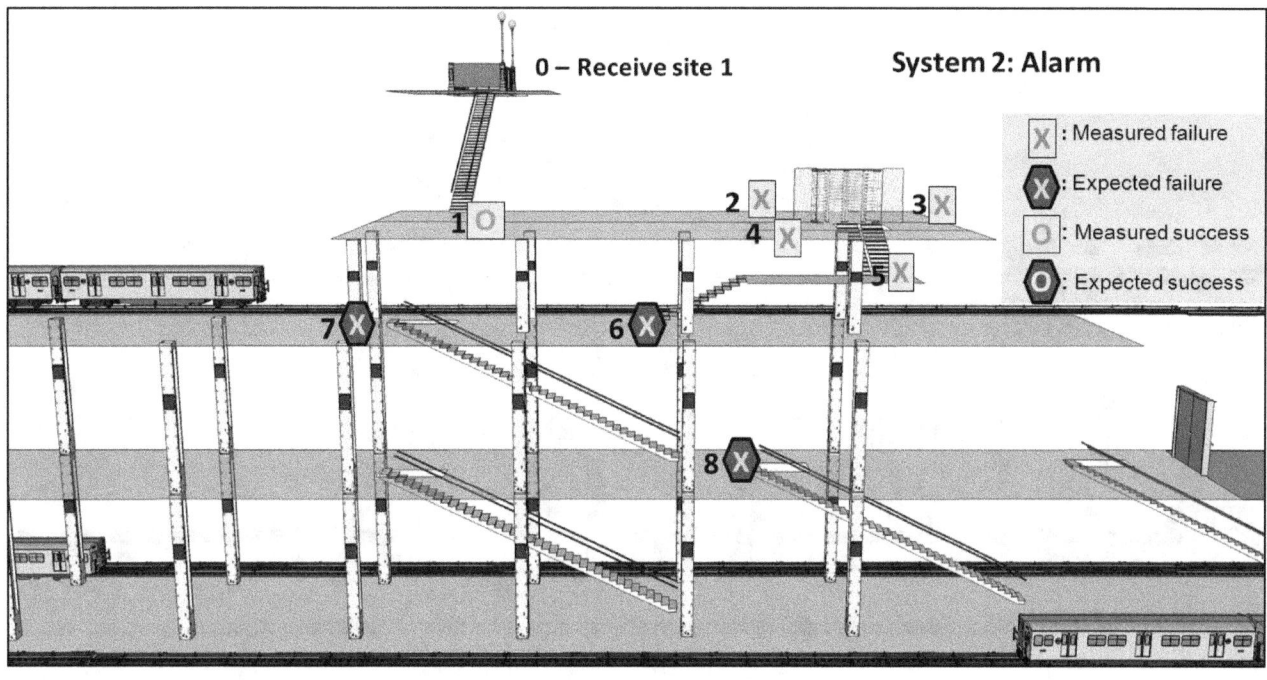

(a)

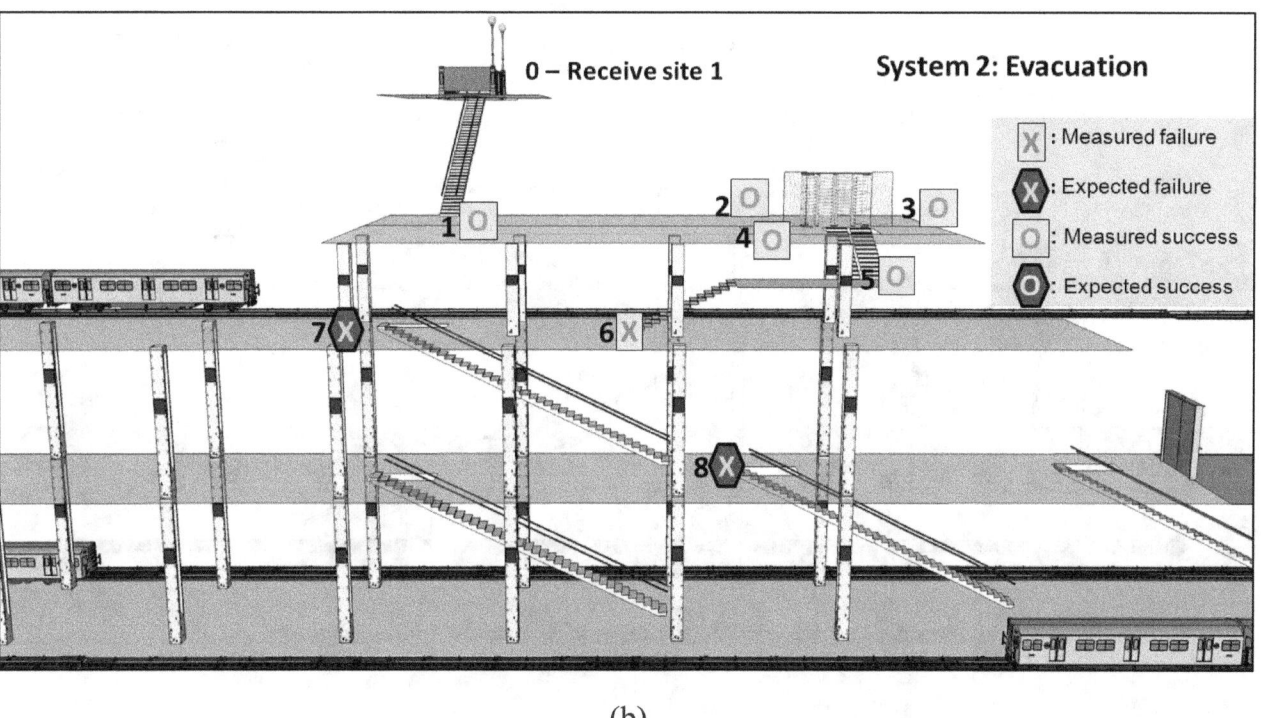

(b)

Figure 13. (a) Firefighter-down "Alarm" signal and (b) "Evacuation" signal for RF PASS System 2. These are results without the repeater.

19

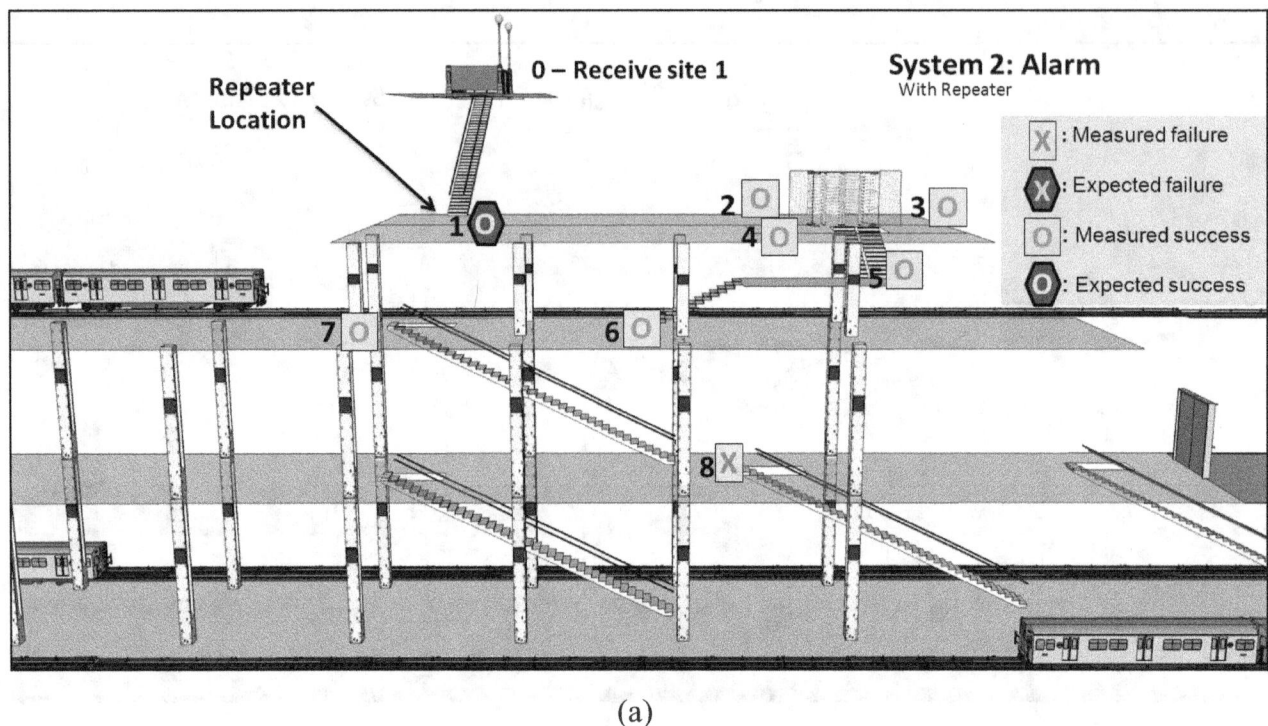

(a)

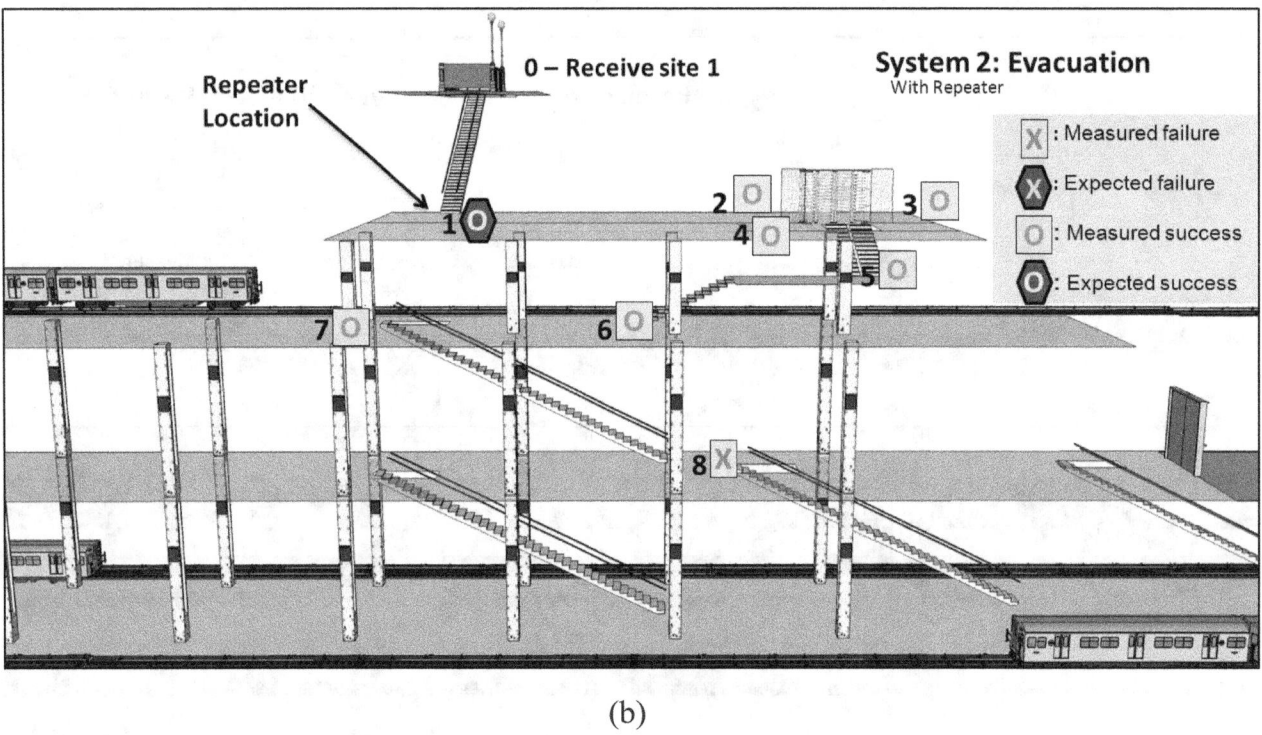

(b)

Figure 14. (a) Firefighter-down "Alarm" signal and (b) "Evacuation" signal for RF PASS System 2, with the repeater unit located at point 1.

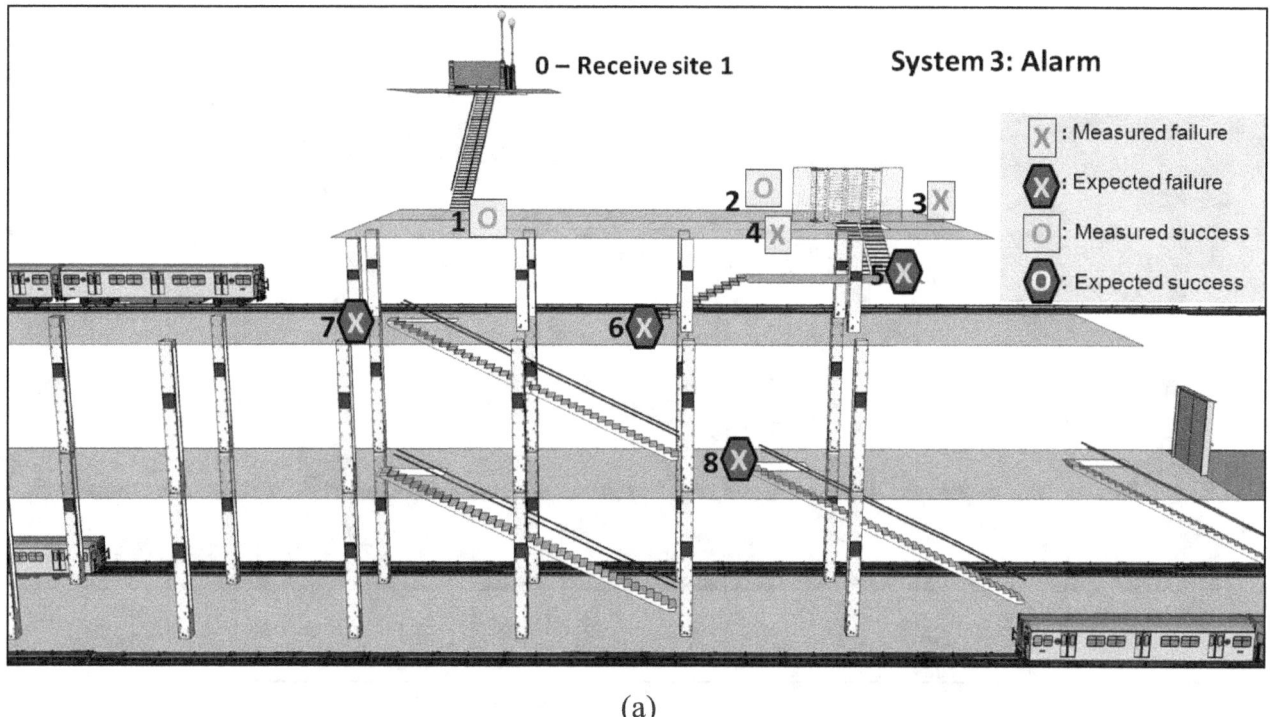

(a)

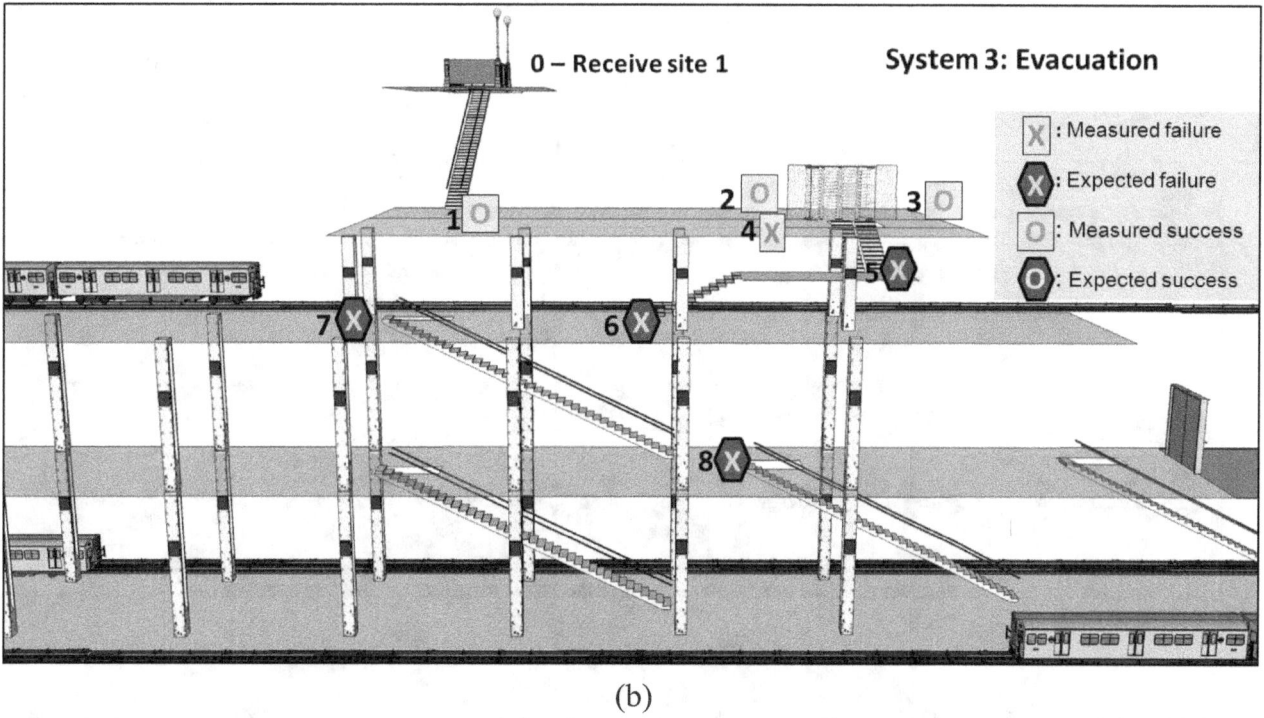

(b)

Figure 15. (a) Firefighter-down "Alarm" signal and (b) "Evacuation" signal for RF PASS System 3. System 3 did not have a repeater option.

21

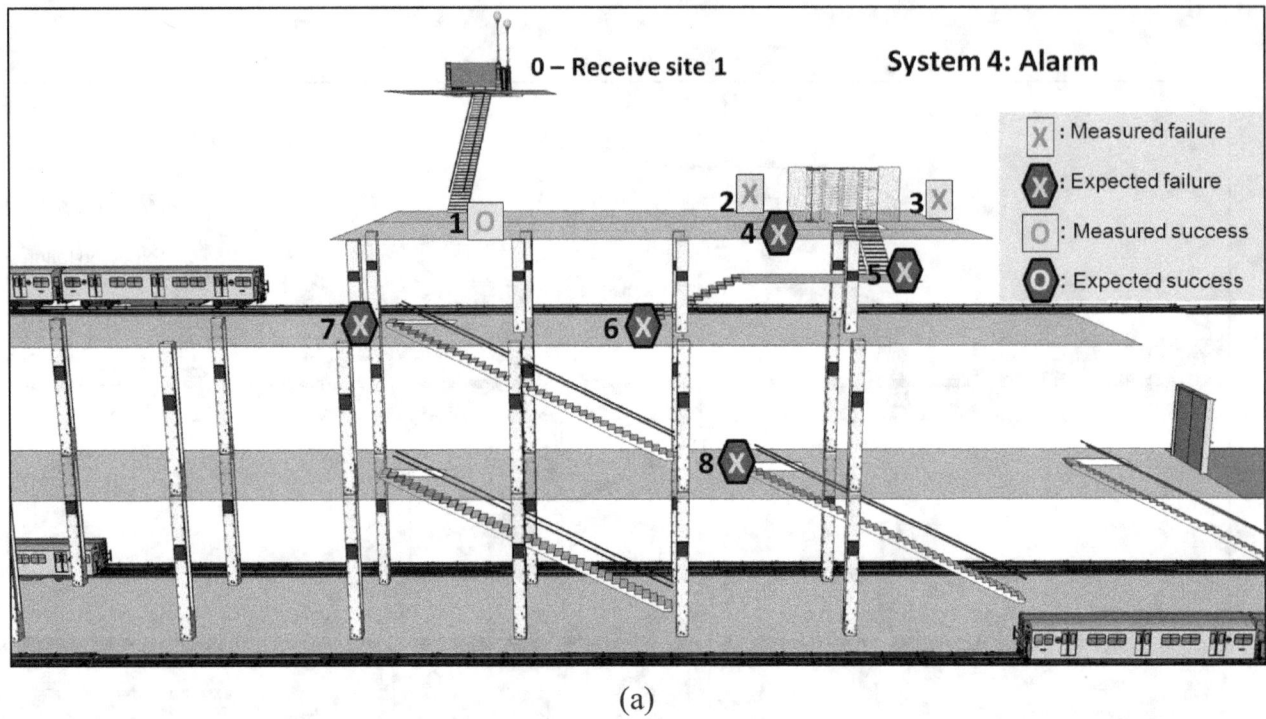

(a)

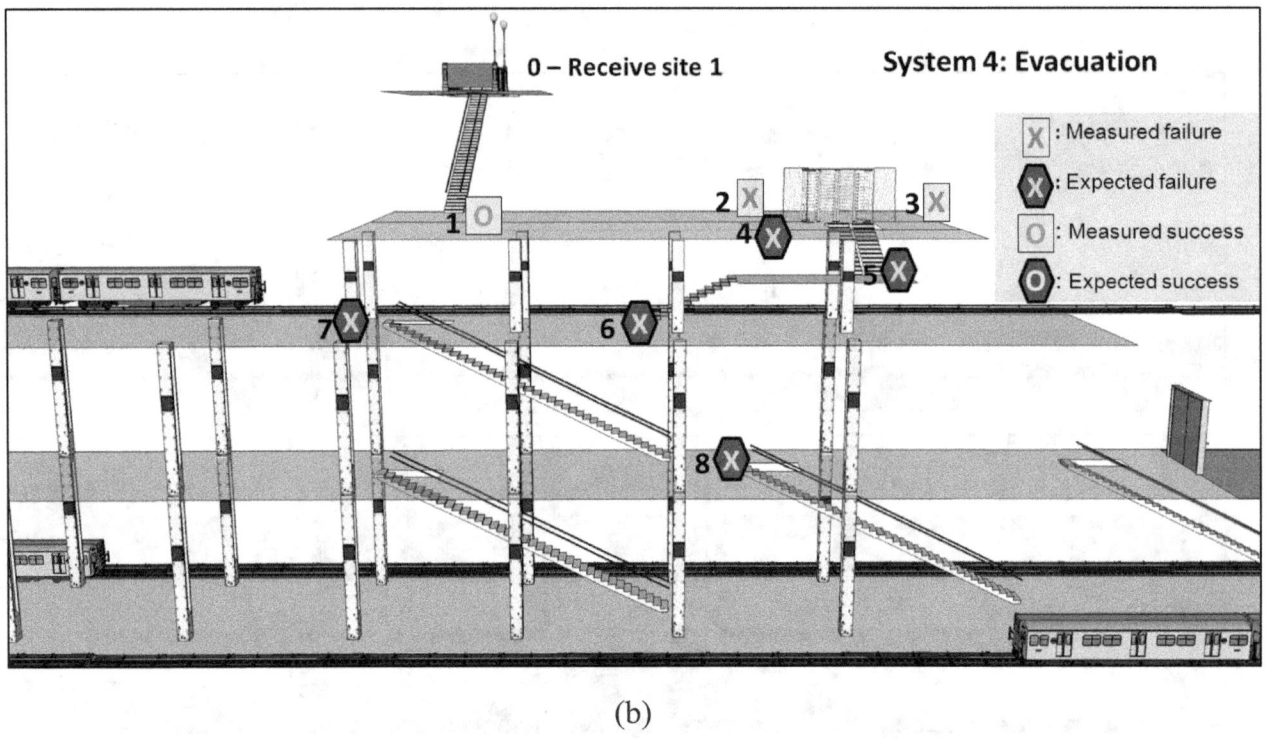

(b)

Figure 16. (a) Firefighter-down "Alarm" signal and (b) "Evacuation" signal results for RF PASS System 4. These results are without a repeater.

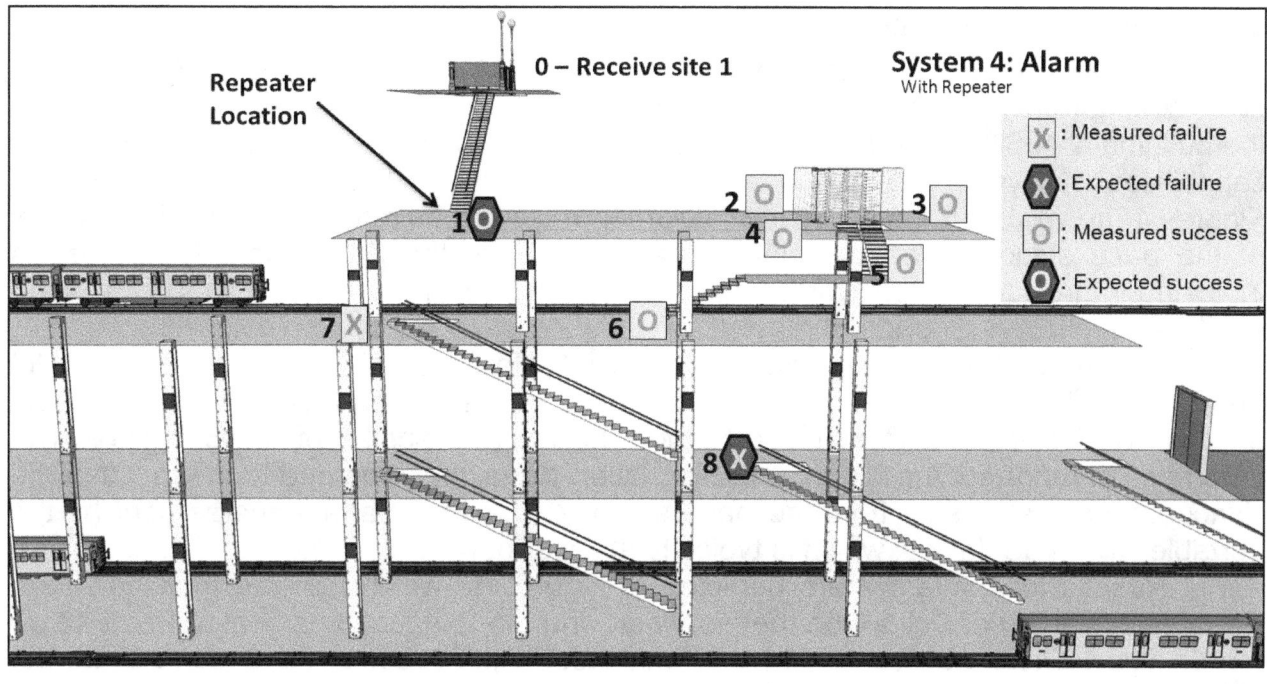

(a)

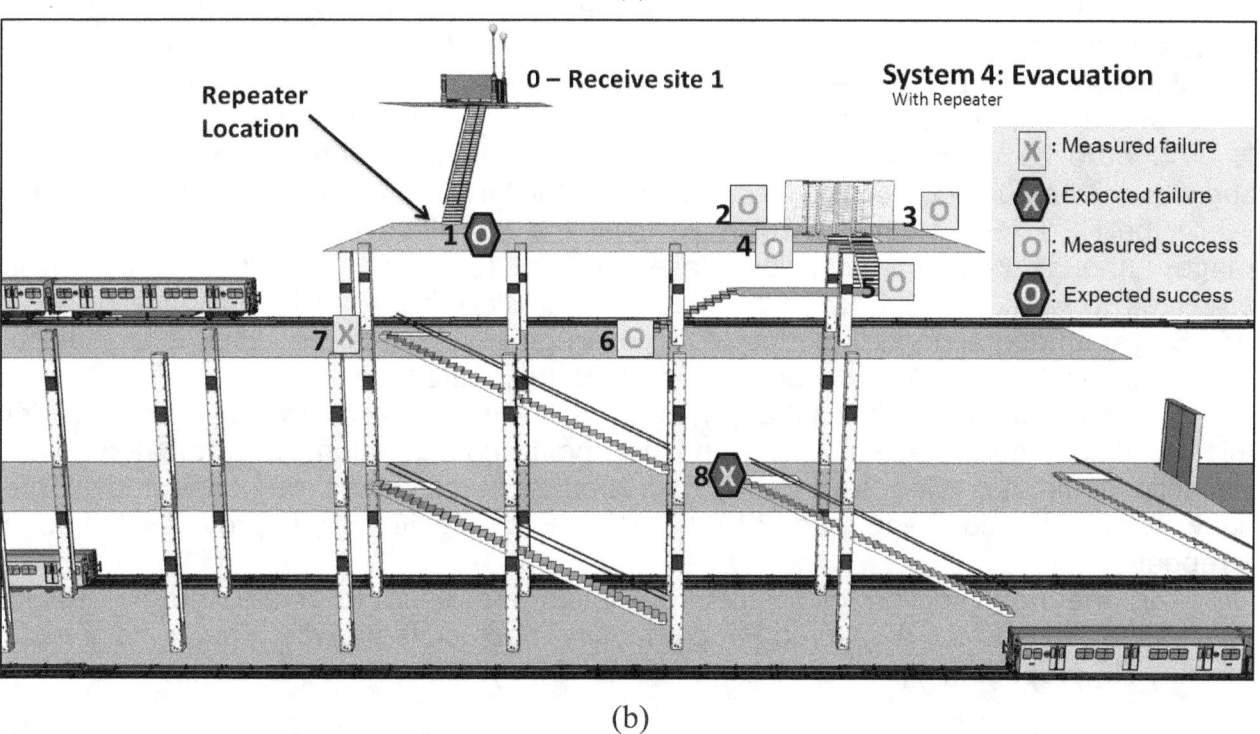

(b)

Figure 17. (a) Firefighter-down "Alarm" signal and (b) "Evacuation" signal results for RF PASS System 4, with the repeater unit located at Point 1.

3.2 Empire State Building

The second set of results covers the four RF PASS systems tested in a point-to-point configuration in the Empire State Building. The base station was located outside the building at NIST Site 1. Due to time restrictions for conducting the tests, the repeater capabilities of Systems 2 and 4 were not tested. As in the previous four figures, Figures 18 to 21 depict the success or failure in receiving a firefighter-down alarm signal at the base station or an evacuation alarm signal at the PASS unit. A success is indicated if the signal is received within 30 seconds; otherwise, the test was considered a failure.

Figure 18 shows test results for System 1. Here, the Alarm signal is successful up to the 29th floor, but fails at the 61st floor and above. The Evacuation signal shows success on Point 24 of the 61st floor and success at Point 29 of the 83rd floor. The difference in range for the Alarm and Evacuation signals may be due to the fact that the Evacuation signal is sent from the base station, while the Alarm signal is sent by the portable RF PASS device, which is typically at lower power.

Results for System 2 are depicted in Figure 19, which indicate that the Alarm signal fails starting at Point 4 on the first floor, and the Evacuation signal at Point 10 on the 5th floor. There was some concern with these Alarm results for System 2, as the software was not resetting properly, and thus, the Alarm results may be indicating a performance less than expected. However, based on the subway performance for the point-to-point configuration of System 2 along with the Evacuation signal results here, the Alarm signal would not be expected to exceed the range of the Evacuation signal. Thus, at best, the range of the Alarm signal would likely extend to parts of the 5th floor.

Figure 20 shows the test results for System 3, where there was demonstrated success on the 20th floor for the Alarm signal at Point 17, but failure at Points 13 and 16. While the RF PASS worked in some parts of the 20th floor, other areas were out-of-range. Although the Evacuation signal failed at Point 13 as well, the Evacuation signal is predicted to work at Point 17. In general, the 20th floor behavior is indicative of a RF PASS system operating at the edge of its coverage limit, where the signal is successful approximately half of the time, but fails the other half of the time.

Finally, Figure 21 shows that System 4 has a very limited coverage capability for this building when operating in a point-to-point configuration. Unfortunately, time restrictions did not allow testing with a repeater because, as demonstrated in the subway tests, the coverage range of System 4 can be significantly improved with use of a repeater.

In the next section of results, bar plots are used to relate the RF PASS success/fail tests to the measured RF path loss. Those plots provide greater insight into this key RF propagation problem that the RF PASS systems must overcome.

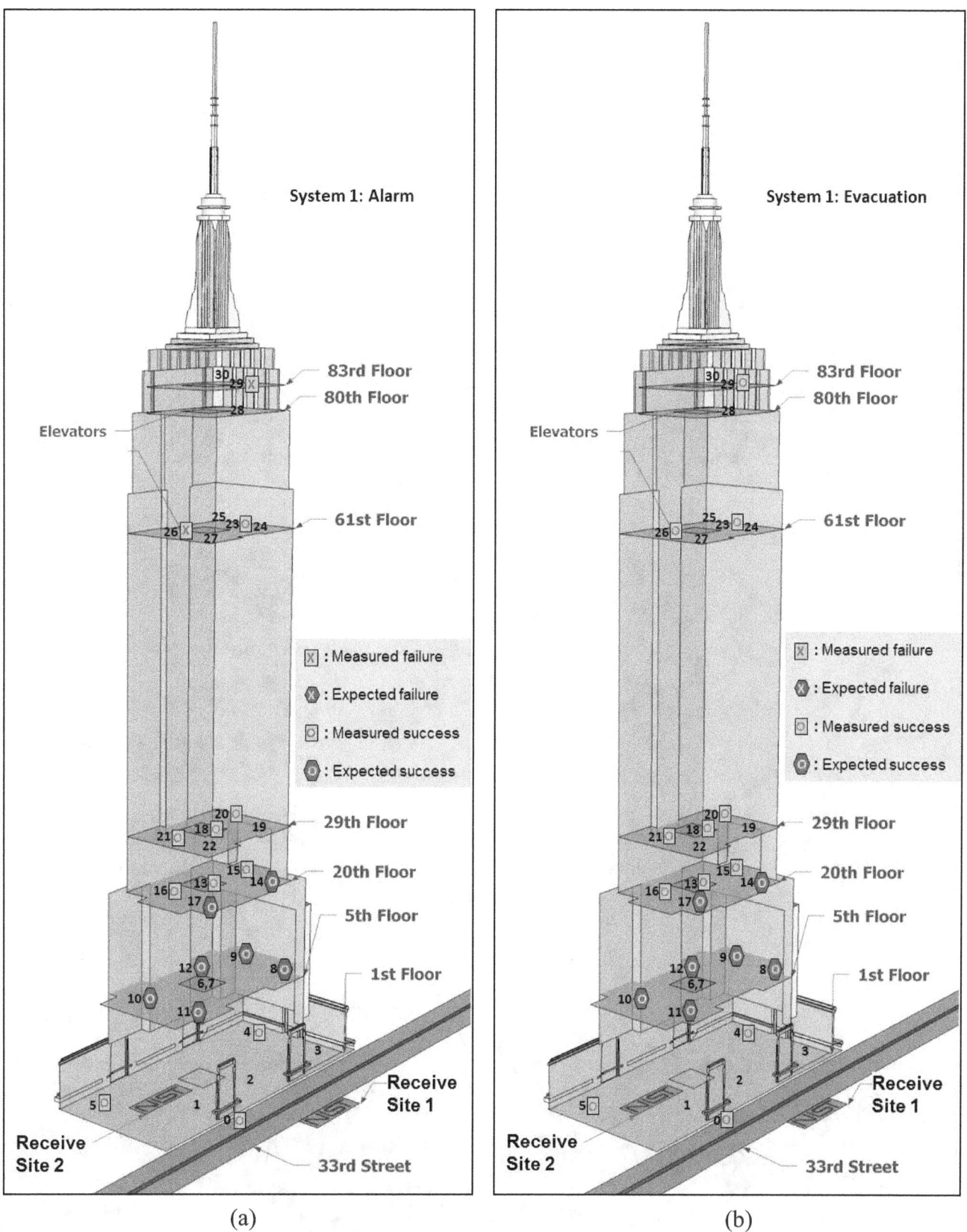

Figure 18. (a) Firefighter-down "Alarm" signal and (b) "Evacuation" signal results for RF PASS System 1 in the Empire State Building.

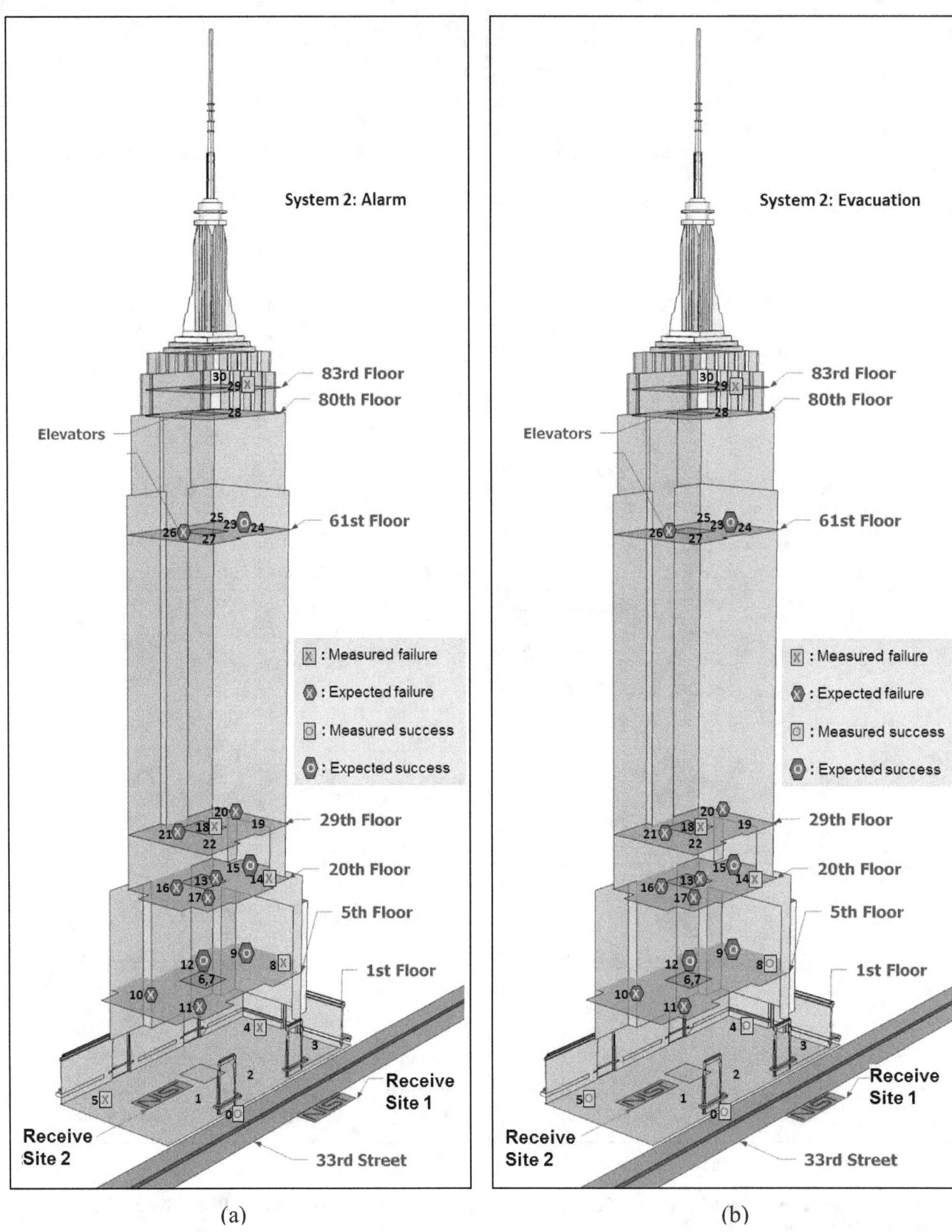

Figure 19. (a) Firefighter-down "Alarm" signal and (b) "Evacuation" signal results for RF PASS System 2 in the Empire State Building.

26

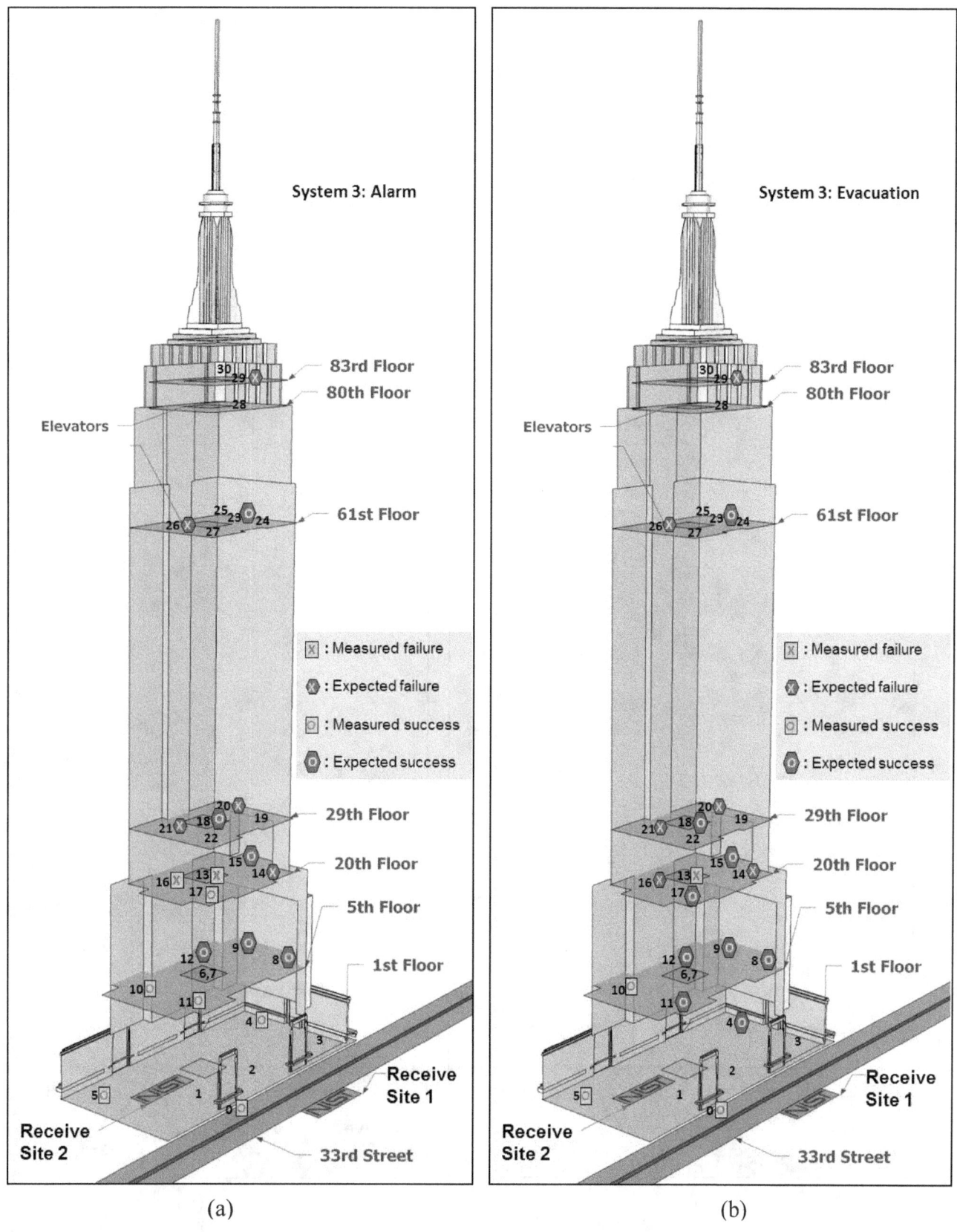

Figure 20. (a) Firefighter-down "Alarm" signal and (b) "Evacuation" signal results for RF PASS System 3 in the Empire State Building.

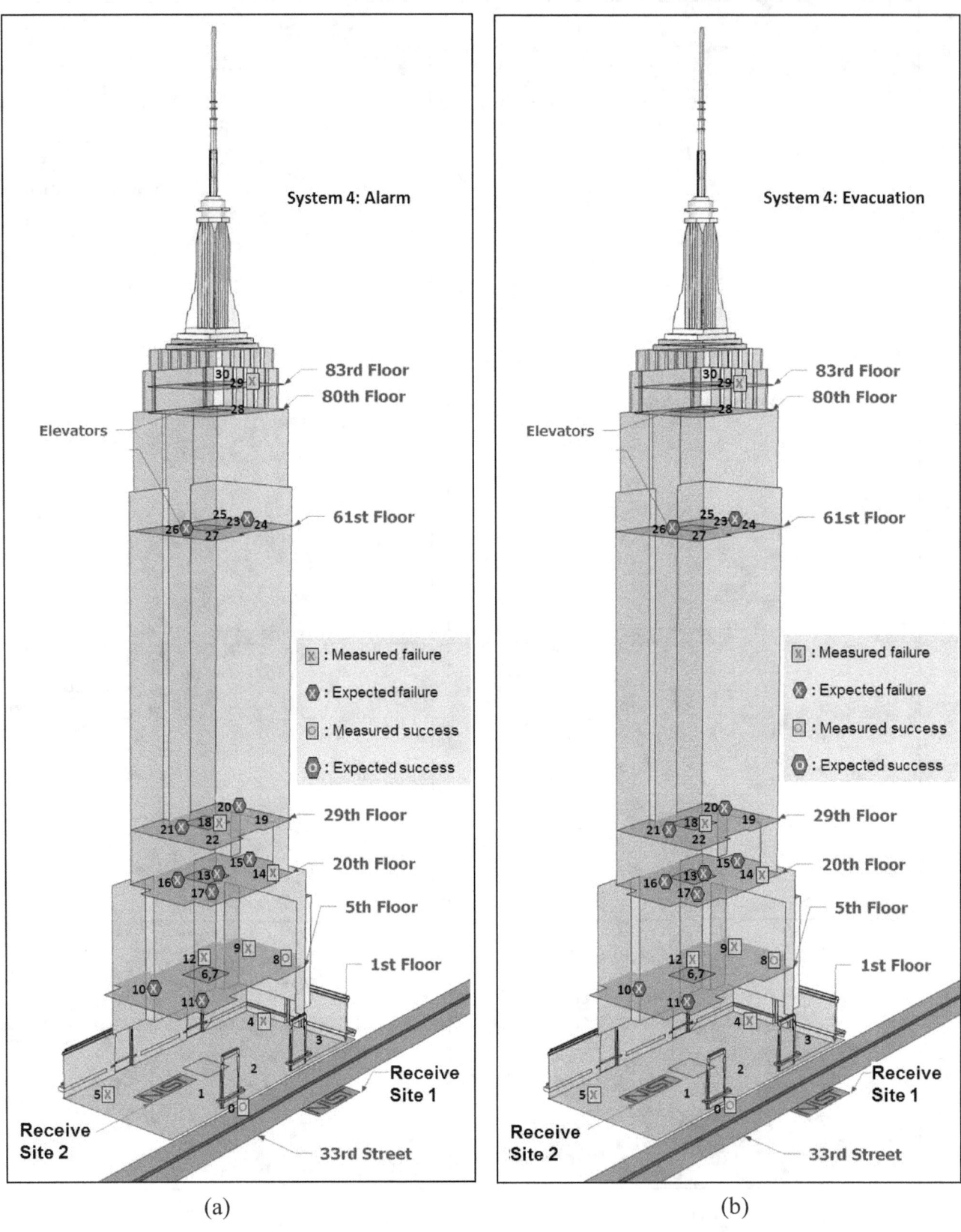

Figure 21. (a) Firefighter-down "Alarm" signal and (b) "Evacuation" signal results for RF PASS System 4 in the Empire State Building.

3.3 PASS system results versus path loss

In this section, we plot the RF PASS test results in Figures 12 to 21 versus the estimated path loss at the test location. The basic approach was to measure the path-loss experienced by a radio operating in the same frequency band as the RF PASS at the same test points. The RF PASS success/failure results for the test point were then associated with estimated path loss value. Sections 4 and 5 discuss the measurement method and data processing used for determining the path-loss. Correlation between the PASS system performances and the path loss at the frequency of radio operations supports the development of laboratory test methods by allowing the accurate re-creation of field-test conditions.

Figures 22 to 28 provide histograms of RF PASS test results versus the measured path-loss at the test location. The success/failure results are placed in bins, 5 dB in width, based on the associated path-loss value. Lines that correspond to an attenuation test threshold of 100 dB (i.e., the NFPA 1982 low-attenuation test) and the effective noise floor of the path-loss measurement system are also shown on the plots. Note that the noise floor limit depends on frequency and, thus, the noise floor lines in the plots correspond to the frequency of operation specific to that RF PASS.

Figure 22 shows the Alarm (or firefighter-down) signal results in the subway for all four systems in a point-to-point configuration. Figure 23 shows the Evacuation signal results for the same configuration. All four systems are able to successfully transmit signals when the path loss is less than the 100 dB threshold. However, between 100 dB and 140 dB of path loss, all systems experience at least one failure in both the Alarm and Evacuation signal tests.

When a repeater is located at Point 1, Systems 2 and 4 exhibit a much better coverage range. Figure 24 shows both the Alarm and Evacuation signal test results for these two systems, where both systems successfully transmit signals with up to approximately 140 dB of path loss. System 4 does not fail until the path loss exceeds 150 dB. There are two failures indicated for System 2 at just over 140 dB, but the path-loss number might actually be closer to 150 dB at that location. This is due to the fact that all of the path-loss numbers are subject to measurement uncertainty and channel variability; these factors will be discussed in Section 7. Also, since the path-loss estimate is near the noise-floor limit of the measurement equipment, distinguishing the signal power from the noise power is difficult. This can cause a lower path-loss estimate than when a sufficient signal-to-noise margin exists. Overall, these single-hop repeater results demonstrate the benefits of the use of repeaters to overcome significant path loss in the communication link.

In Figures 25 and 26, the Alarm and Evacuation results are shown for the test carried out in the Empire State Building. Systems 1 and 3 both successfully transmit Alarm and Evacuation signals with more than 100 dB of path loss. System 1 demonstrates the ability to successfully transmit most signals near the 130 dB noise floor of the measurement system, while System 3 transmits with success up to approximately 105 dB to 110 dB of path loss.

System 2 generally transmits successfully beyond the 100 dB threshold, and exhibits better performance for the Evacuation signal than the Alarm signal. To some extent, the difference between the two signals for System 2 is consistent with the subway results. However, as mentioned previously, there appeared to be some reset issues with the software when performing the Alarm tests, so those results should be viewed with some skepticism. The Evacuation results, along with the previous subway results, indicate that the system can overcome a 100 dB of path loss without a repeater.

System 4 indicates a failure of both the Alarm and Evacuation signals for 100 dB of path-loss. However, the estimated path loss value bin is 5 dB wide, and thus represents a 97.5 dB to 102.5 dB range. Thus, the actual path-loss value may be higher than 100 dB, which is likely the case at this test point; e.g., the path-loss could be 102 dB at this test point, but the result falls in the 97.5 to 102.5 dB bin. Again, as mentioned before, there is uncertainty is the measurement of the path-loss values as well. These uncertainties will be discussed in Section 7.

Figures 27 and 28 contain the aggregate results (without repeaters) from the subway station and the Empire State Building. One of the important factors to notice is that the path-loss values are lowest for System 1 and highest for System 4. This is due to a difference in operating frequency as the systems were tested at the same points in the buildings. The frequencies of operation for the tested RF PASS systems ranged from the 400 MHz to the 2.4 GHz bands. As will be discussed in Section 4, the path-loss values are based on continuous-wave transmitters, and provide a good estimate across the operational bands of the RF PASS.

Except for a few cases, the Alarm and Evacuation signals for System 1 are successful up to 120 dB of path loss. System 2 successfully communicates the Evacuation signal with 100 dB of path loss, and as mentioned previously, the failure of the Alarm signal around the 100 dB threshold is likely attributed to a software reset issue. System 3 successfully communicates Alarm and Evacuation signals up to 105 dB in all except a few cases around 105 dB. In the case of System 4, a majority of the test points exceed 100 dB of path loss, with many test points experiencing greater than 125 dB of path loss. Without a repeater, System 4 is not able to overcome path loss-values of this magnitude. Note that none of the other point-to-point systems consistently and successfully communicated Alarm and Evacuation signals when the path-loss level was close to 125 dB.

Finally, the estimated path-loss values associated each test point, along with the success/failure results are presented in Appendix A. Tables 7 and 8 provide the results for the Alarm and Evacuation for the subway station and the Empire State Building, respectively. The path-loss values listed in these two tables are determined as described in Section 4 below.

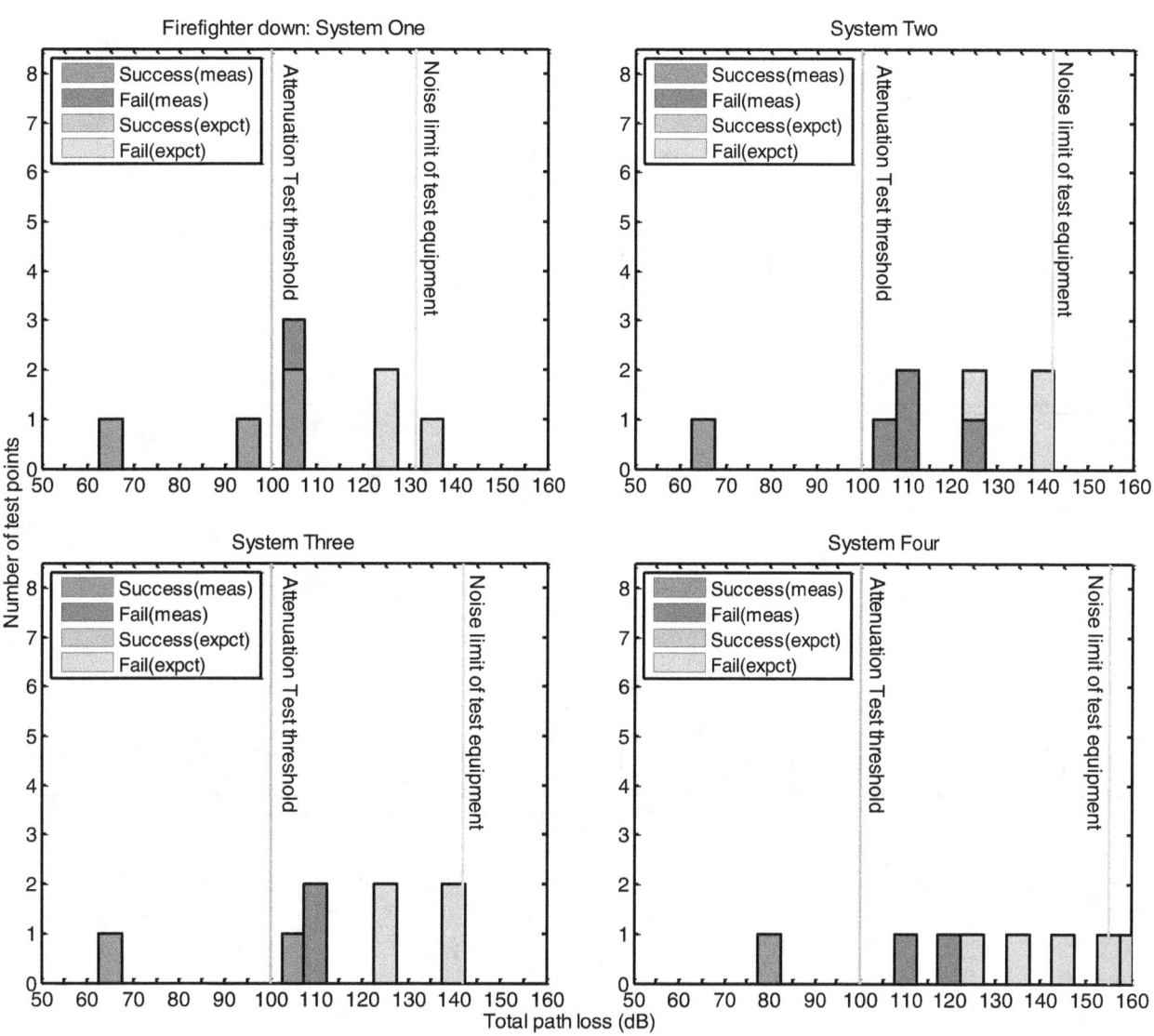

Figure 22. Subway firefighter-down signal results without a repeater. Plots are based on data contained in Table 7.

31

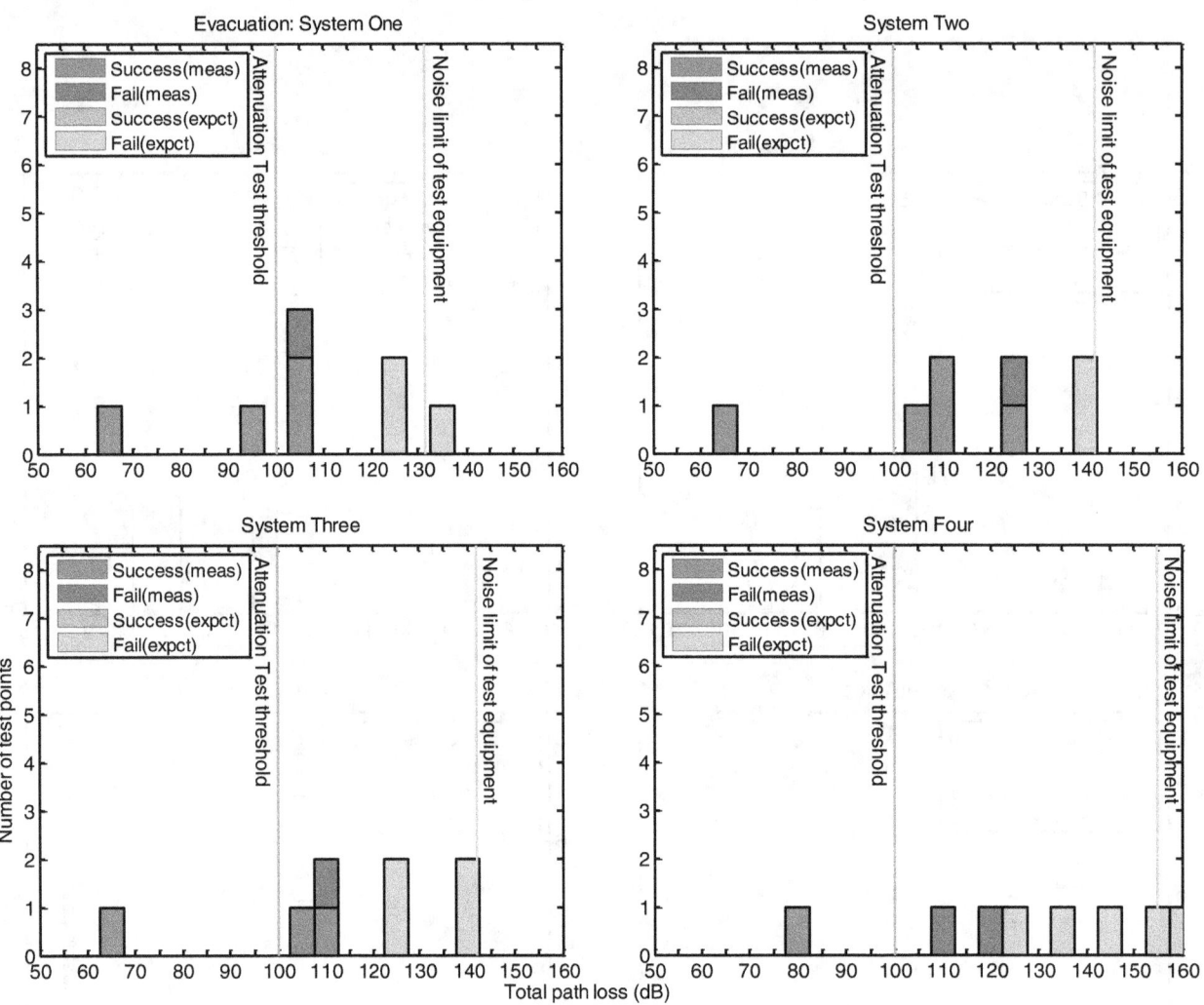

Figure 23. Subway evacuation signal results without a repeater. The Attenuation Test threshold is 100 dB for all systems, and represents the NFPA Standard. The noise limit depends on the test equipment and is frequency dependent. Plots are based on data contained in Table 7.

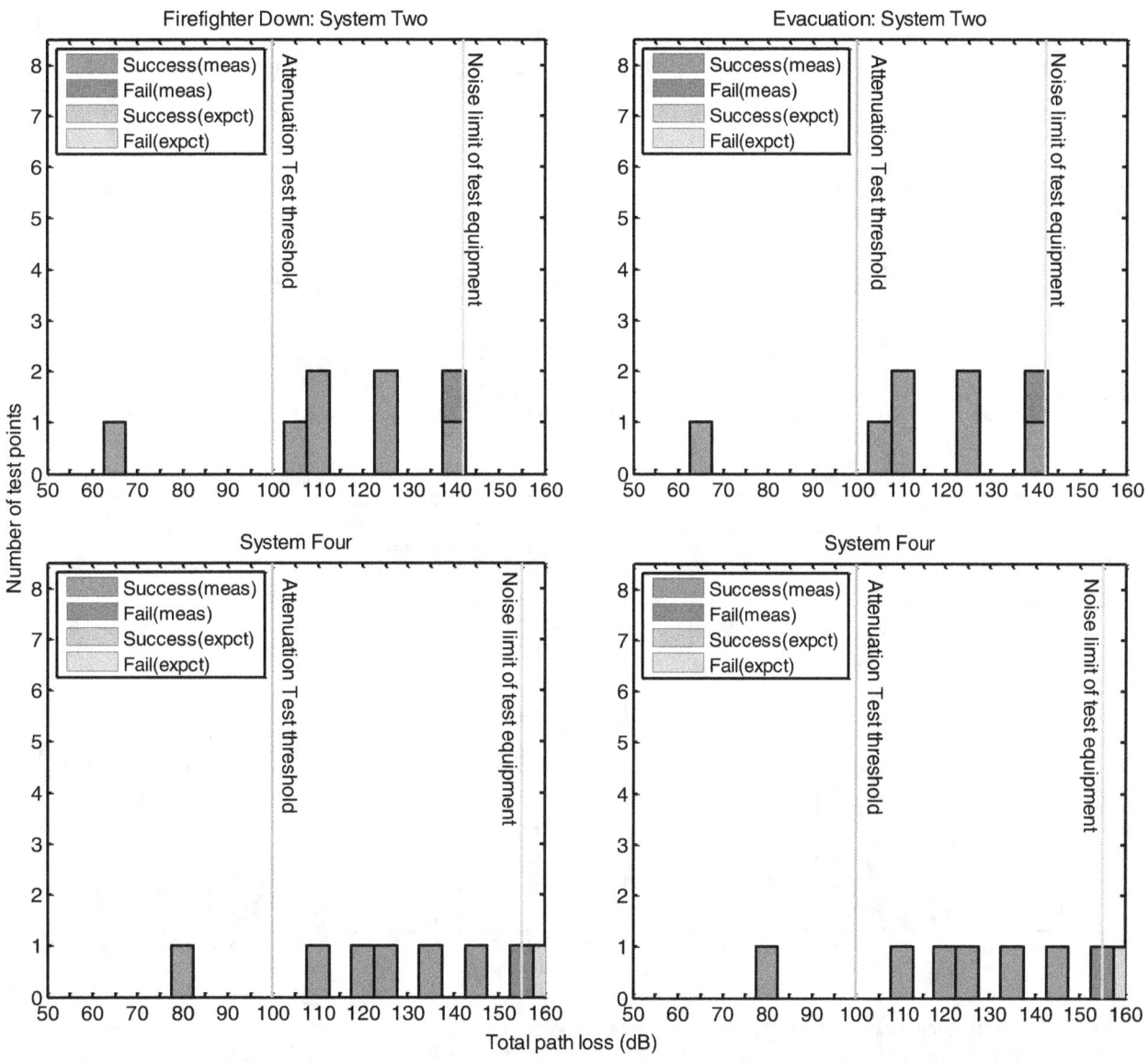

Figure 24. Subway firefighter-down and evacuation results with a single-hop repeater. Only Systems Two and Four were tested with a repeater node. Plots are based on data contained in Table 7.

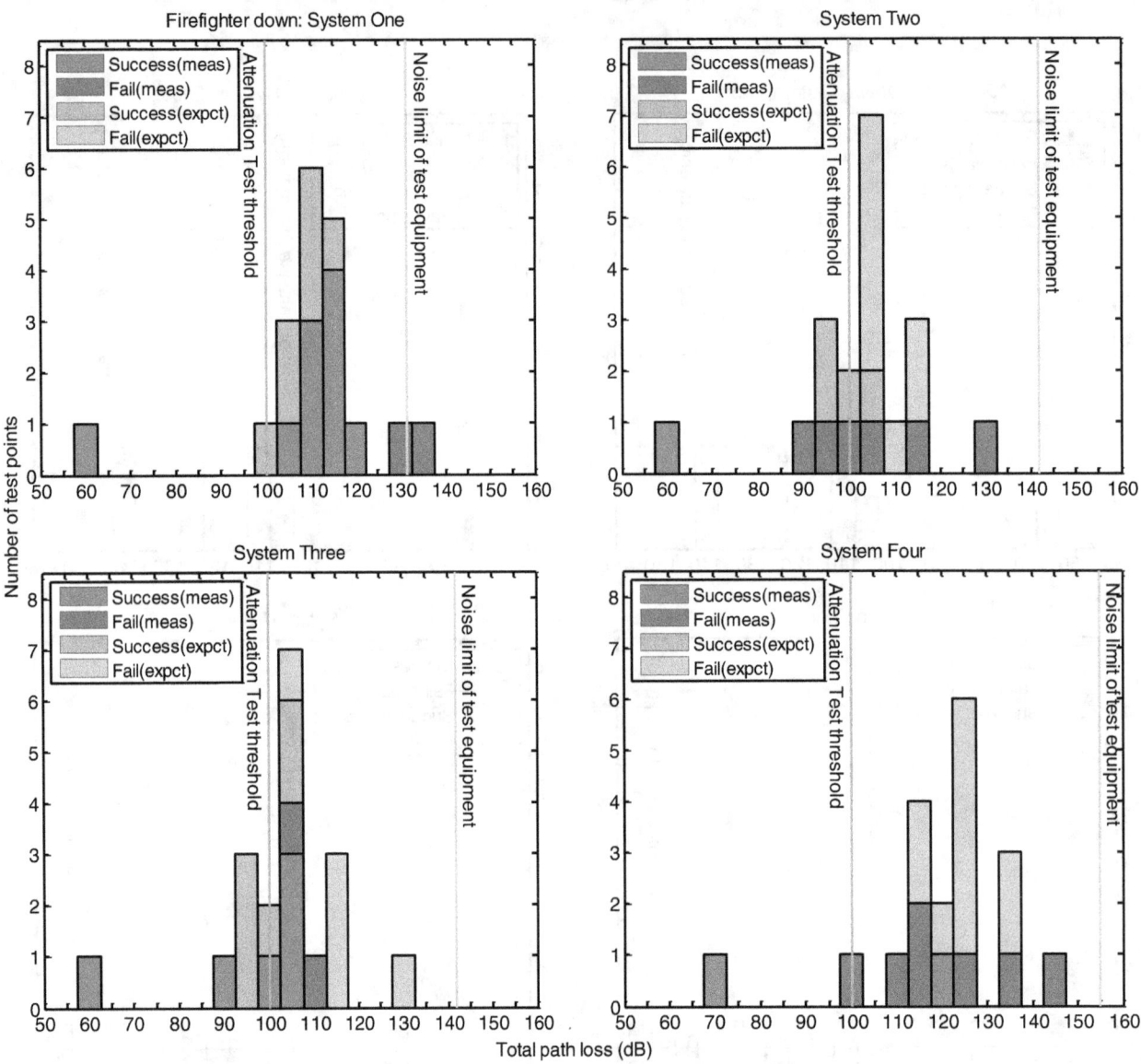

Figure 25. Empire State Building firefighter-down results without a repeater. Note that the software for System 2 did not appear to be resetting properly during the test. Thus, the results for firefighter-down alarm for the Empire State Building should be viewed with skepticism. The cause of the problem was not identified, but the software for System 2 had not previously shown a reset problem over the last several years of our testing in the field. (RF interference did not appear to cause the problem.) The evacuation signal did not appear to be impacted. Plots are based on data contained in Table 8.

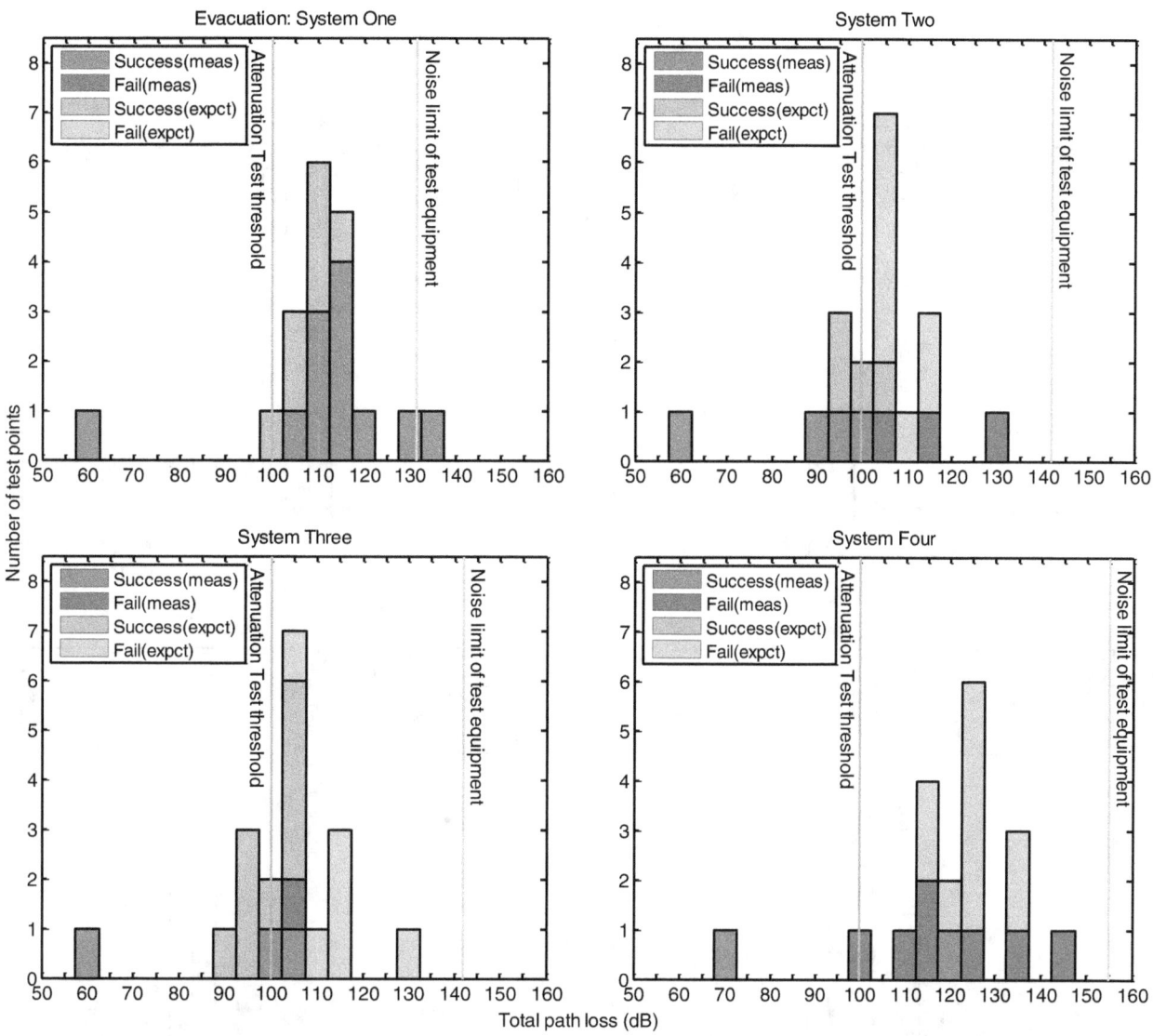

Figure 26. Empire State Building evacuation signal results without a repeater. Plots are based on data contained in Table 8.

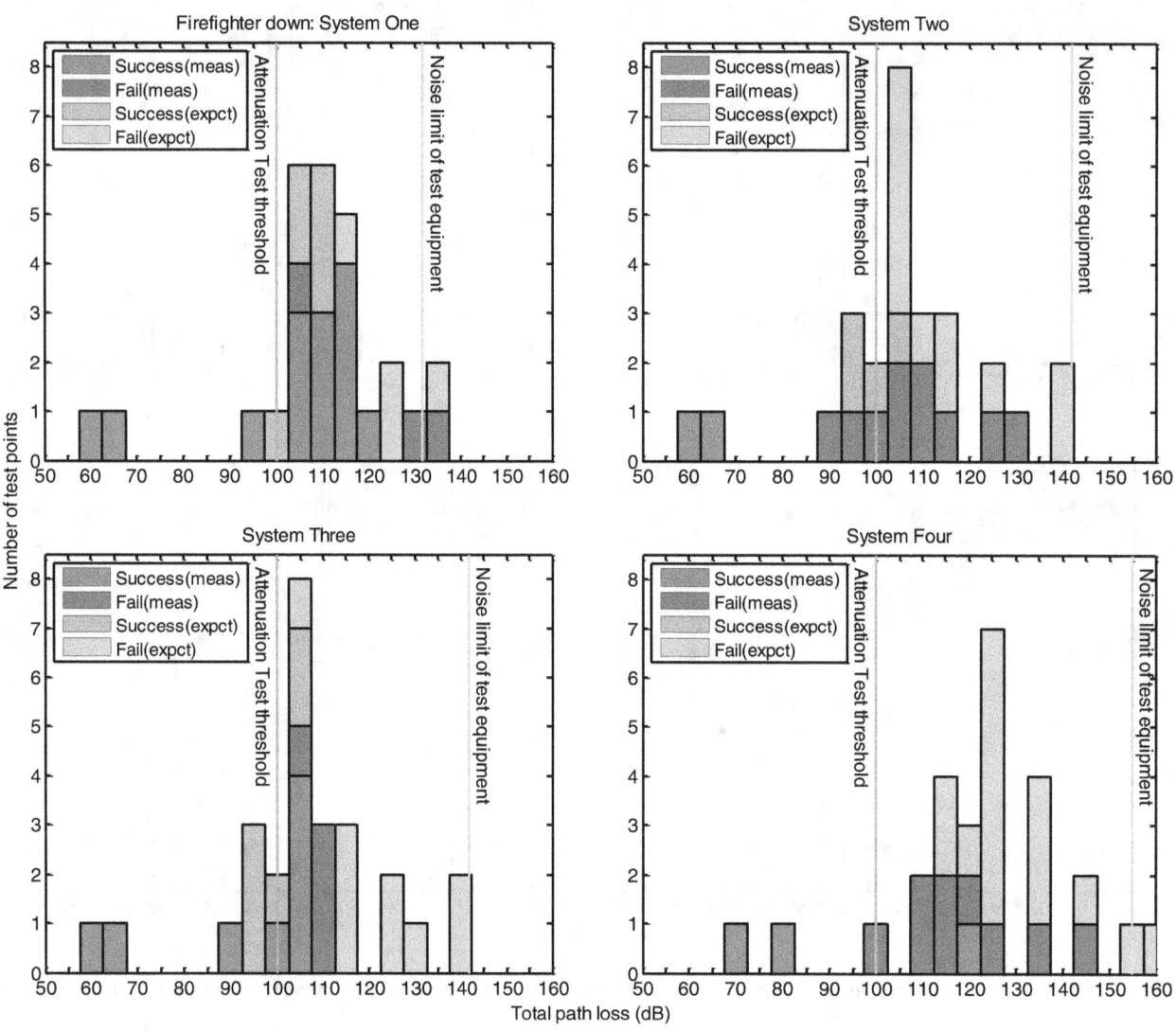

Figure 27. Aggregate firefighter-down results for the subway and Empire State Building tests without a repeater. Plots are based on data contained in Table 7 and 8.

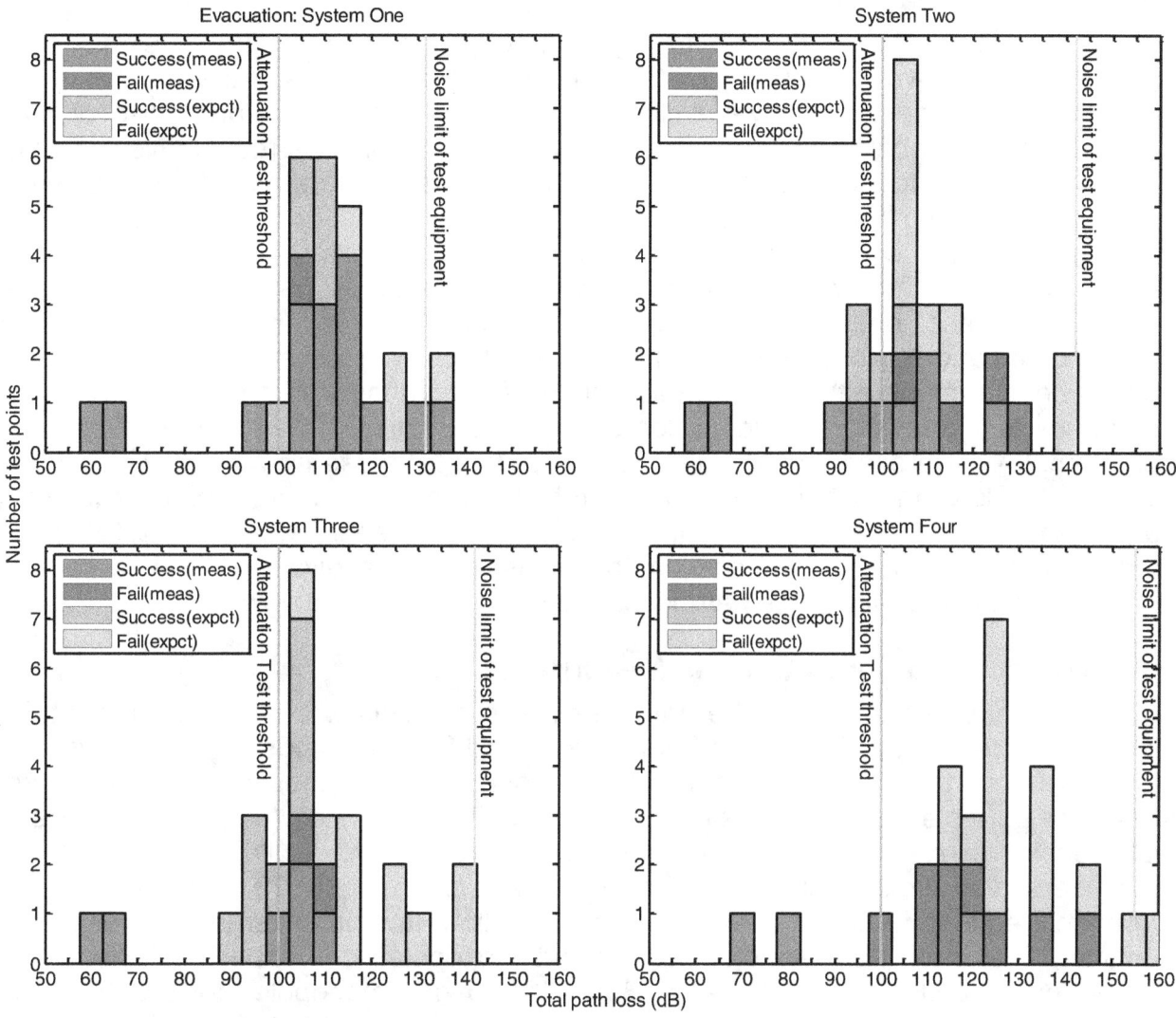

Figure 28. Aggregate evacuation signal results for the subway and Empire State Building tests without a repeater. Plots are based on data contained in Table 7 and 8.

4. Path-Loss Measurement

The path-loss, or reduction in signal strength experienced by a signal as it penetrates and travels through a structure, will directly impact the ability of an emergency responder to receive a signal from the incident command post, or vice versa. The primary goal of the measurements was to determine representative values of this key impairment to successful transmission of alarm signals from RF PASS systems. The basic approach taken to obtain these path-loss measurements used a radio-mapping technique, where a continuous-wave (CW) transmitter is carried throughout the building while a receiver located elsewhere captures the signal. Typically, as is the case here, at least one receive location corresponds to location that simulates the incident command station. Additional receive locations within the two structures also provide measurement of the signal propagation as the transmitter is carried throughout the various tunnel levels or building floors. The combination of data from the external and internal receive sites enables us to estimate the path loss from the street level to the lower levels of the subway station. In order to correlate the RF PASS results with the propagation measurements, the transmitter is held at each RF PASS test location for a short period of time. Details on the transmitters, receivers, and data-capture process used in determining the path-loss values are discussed below.

4.1 Transmitters for Path-Loss Measurements

The transmitters used in the path-loss measurements were chosen to meet the following criteria: (1) transmit at the frequencies used for the various RF PASS systems and other frequencies of interest in Table 1, (2) operate continuously for several hours, and (3) be portable in a manner similar to the RF PASS devices (i.e., a single person could easily carry the device while walking through the building.) Commercial transmitters in plastic protective cases, such as depicted in Figure 29, were available for all of the frequency bands. These transmitters allow for the external connection of antennas. The antennas were electrically short monopoles or "rubber duck" types for the frequency bands of 430 MHz and 1834 MHz, loaded monopoles for the 750 MHz and 905 MHz bands, and a linear dipole-array antenna for 2.4 GHz. All of these antennas generate an omnidirectional pattern, with nominal gains ranging from 0 dBi for the rubber duck to 5 dBi for the linear dipole-array.

Table 1. Public-safety communication frequencies and cellular bands, including nominal frequencies used in the measurements.

Frequency Band (MHz)	Nominal frequency used in measurements	"Transmit" antenna	Description
406-470	430	electrically short monopole or "rubber duck"	Used by public-safety and others, including RF PASS technology
700-800	745,750	loaded monopole	Includes the new public-safety band
902-928	905	loaded monopole	Unlicensed band; used in RF PASS technology
1850-1990	1834	electrically short monopole or "rubber duck"	PCS or digital systems
2400-2500	2405	linear dipole array	Unlicensed band; wireless LANs; used in RF PASS technology

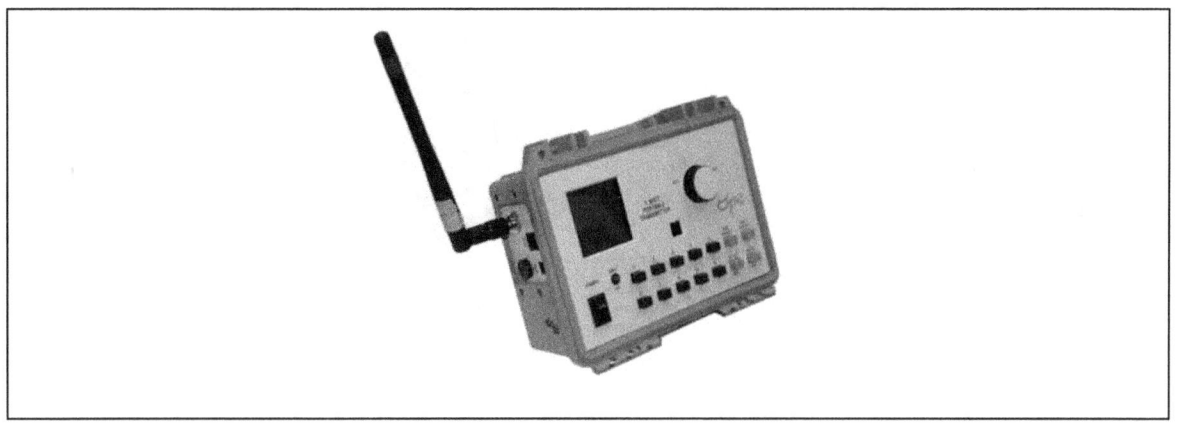

Figure 29. Typical transmitter with a "rubber duck" antenna used to generate the CW signal

The 430, 750, 905, and 1850 MHz radios had nominal output powers of 1 W (30 dBm), and the nominal output power of the 2405 MHz radio was 5 W (37 dBm). However, rather than work with nominal transmitter power levels, we opted to measure the total radiated power (TRP) of the various transmitter/antenna combinations so as to calibrate the transmitted power in the path-loss calculations. The measured TRP values were based on reverberation chamber measurements. The benefit to this approach is evident by the fact that the TRP values never reach the nominal values due to factors such as antenna efficiencies, cable losses (the cable connecting the antenna to the transmitter), and connector losses. Measured TRP values also account for antenna-to-radio impedance mismatches, which can significantly impact the actual radiated power. These measured TRP values are incorporated into the calibration process discussed in Section 4.3.

4.2 Receiver for Path Loss Measurements

As shown in Figure 30, the receiver consisted of a spectrum analyzer controlled by a graphical programming language. The software ran parallel processes of collecting, processing, and saving the data for post-collection processing. The data were continuously read from the spectrum analyzer and stored in data buffers. The software processed these buffers by searching for the power level of the desired CW signal, which was subsequently displayed for the operator. These processed data were then saved (along with the original raw data) on the laptop containing the control software.

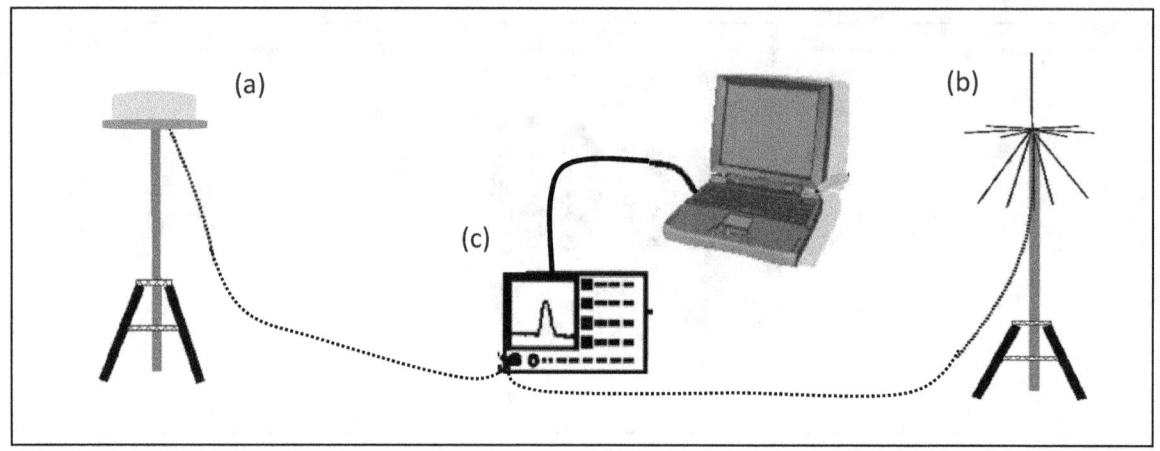

Figure 30. Receiver equipment and setup. The choice of receive antenna depends on the frequency band: (a) conical antenna for 1.8 GHz or a linear dipole-array antenna for 2.4 GHz, and (b) 50 MHz to 1 GHz discone antenna. (c) The spectrum analyzer was connected to a laptop via a GPIB interface.

The sampling rate of the complete measurement process and the walking speed determined the spatial resolution during radio-mapping experiments and the time resolution for recording the signals. A narrow frequency band (called the capture bandwidth) around the desired CW signal was measured during each walk, resulting in sampling period of between 0.15 s and 0.3 s. Continuous-wave sources (i.e., the radios) were carried through the buildings and the receiver recorded the capture bandwidth around the CW frequency. The capture bandwidth was typically less than 20 kHz, but wide enough to ensure that the CW signal was measured even if the CW source experienced frequency drift.

Linear dipole arrays with gains ranging from 3 dBi to 5 dBi were used as the receive antenna in Figure 30 for the 2.4 GHz frequency band. A discone antenna was used for the frequency bands below 1 GHz, with a gain of 2 dBi and a beamwidth of approximately 45 degrees. Finally, a conical antenna was used for the 1.8 GHz receive antenna, with a gain of approximately 3 dBi. When the receive site was outside the structure, the receive antenna was mounted with a short tripod on the top of a van or a fire-emergency medical vehicle for a total height of approximately 4 m above the ground. At receive sites inside the structure, the receive antennas were located on a tripod at a height of approximately 2 m above the ground. Figure 31 depicts the complete transmitter and receiver setup, and indicates the typical orientation of the CW transmitter (Figure 31(a)) and the receive antenna location (Figure 31(b)) for an external receive site.

 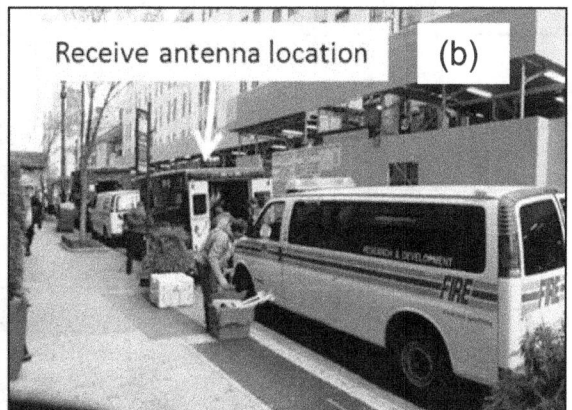

Figure 31. Transmitter and receiver details, (a) a CW transmitter carried through the building and (b) the receive antenna location on an emergency vehicle.

4.3 Data Normalization

A number of post-processing steps were necessary to "normalize" the received-power level of the measured data. These normalized data were then used to obtain path-loss values corresponding to RF PASS test positions and the path-loss for various floors and levels of the structures. These steps are detailed below.

The first step began with a refined power-level search, where the software searched for the maximum signal within the capture bandwidth (between 10 to 20 kHz) by use of a smaller search bandwidth (less than 10 kHz). This search bandwidth was set even smaller, less than 5 kHz, if an interference source lay within the original capture bandwidth, and the CW source did not experience significant frequency drift. The maximum value was obtained over a given time interval. The use of narrower search bandwidth allowed us the flexibility to remove nearby (in frequency) interference in the measured data.

As stated earlier, the radios from 430 MHz to 1.8 GHz nominally transmitted 1 W, while the 2.4 GHz radio transmitted 5 W. To compare received signals, the measured signal power was reduced by the difference in transmit power, approximately 7 dB between 1 W and 5 W, by subtracting 7 dB from the measured 2.4 GHz path-loss data. These path-loss data were also corrected for the measured total radiated power (TRP) of the transmitter/antenna combination to allow comparison across the range of frequencies (see Figure 32). The TRP measurements provide a correction factor that ranges from 1.7 dB to 8.5 dB for the transmitter/antenna combinations used here.

41

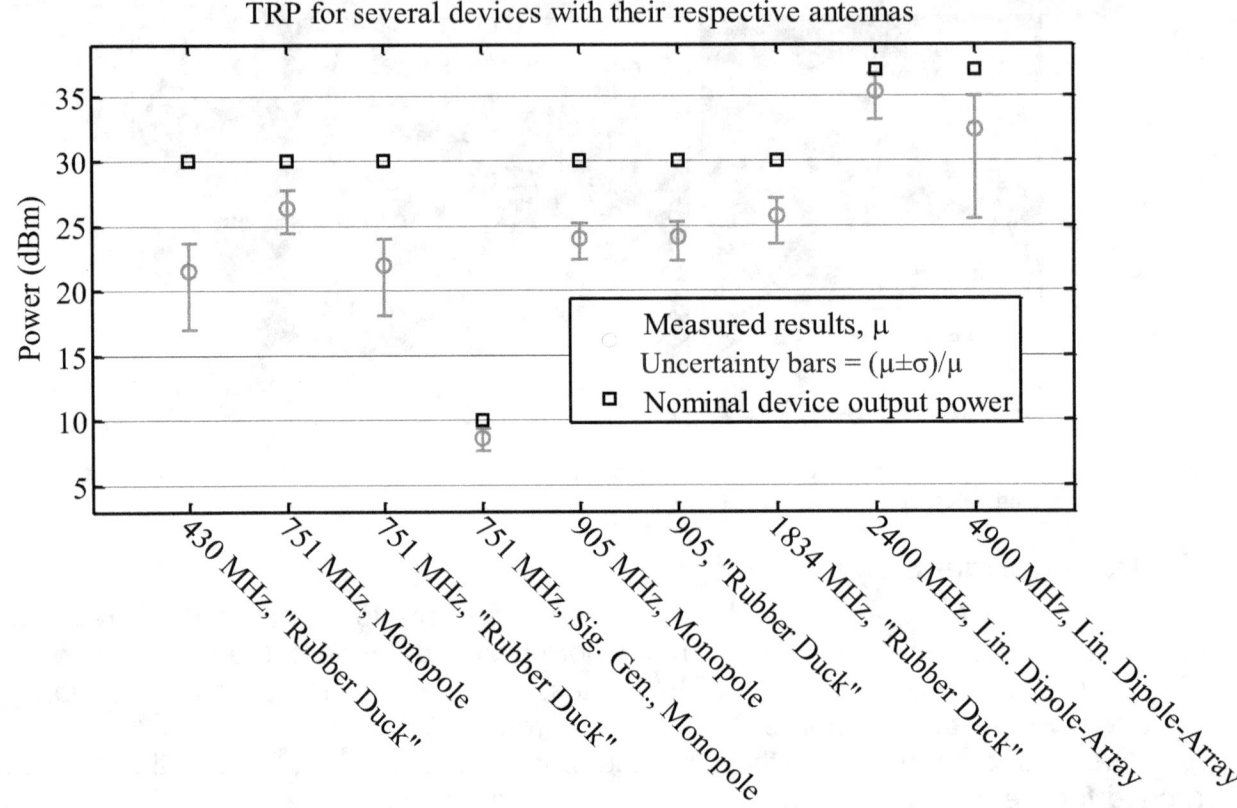

Figure 32. Total radiated power measurements of the CW radios used in the measurement of the path loss. The μ and σ values refer to the mean and the standard deviation as measure in a reverberation chamber at NIST.

The normalized power is then given by

$$P_{RX}^{norm}(\text{dBW}) = [P_{RX}^{raw}(\text{dBm}) - 30(\text{dB})] + \Delta_{TRP}(\text{dB}) - P_{TX}(\text{dBW}) - G_{RX}(\text{dB}), \quad (1)$$

where P_{RX}^{norm} is the normalized received power, P_{RX}^{raw} is the measured received power before any normalization, Δ_{TRP} is the difference from the nominal transmitted power, P_{TX} is the nominal transmitted power, and G_{RX} is the gain of the receive antenna. The gain of the transmit antenna is not included because the electromagnetic scattering environment of the interior of the building reduces the benefits of antenna directivity; the transmit antenna efficiency and mismatch are included in the Δ_{TRP} term. The 30 dB subtraction in the square bracket converts the received power from dBm to dBW. Note that the transmit power is given relative to 1 watt, which is 0 dBW and 7 dBW for the 1 W and 5 W transmitters, respectively. To a first approximation, the receive system (i.e., spectrum analyzer, cables, and receive antenna) and the antennas on the transmitting radios are assumed to impact the measured power in a manner roughly equivalent to an actual deployed public-safety radio network or other wireless system.

5. Path-Loss Behavior for Various Floors of the Structures

In this section, we discuss methods to visualize and analyze the collected data on several of the floors in the Empire State Building and the four levels within the subway station. (All of the data considered in this section are the normalized data described in the previous section.) In the first representation, the data are presented as empirical cumulative distribution functions (ECDFs) with corresponding log-normal distribution curve fits to data measured at individual floors or levels. Parameters for the log-normal curves are provided as well. The ECDF is a direct representation of the data that shows what percentage of the data is below a particular power level. No assumptions about the underlying distribution are required to interpret the ECDF. A log-normal fit provides the parametric values for generating representative distributions that can be used in channel simulations. The log-normal distribution represents the typical power distribution associated with the large-scale fading that is anticipated in office building and urban environments [11]. In addition, as discussed in [12], the log-normal distribution is well-suited for a variety of data analysis problems, several of which exhibit behaviors similar to the data sets collected here. In general, the log-normal distribution provides a good representation of these data, as the data cannot be less than zero and have only a few relatively large values [13].

In the second data presentation, path-loss information derived from the data is displayed in the form of box plots, including the 25, 50, and 75 % quartiles. Box plots are a non-parametric method of displaying data that illustrate the basic symmetry and spread of the data [14]. A key benefit to use of a box plot is that data are not assumed to follow an underlying distribution; e.g., Gaussian or Rayleigh. The values at the 25, 50, and 75 % quartiles are also listed in accompanying tables.

These data representations provide insight into the propagation behavior of the individual floors. Additional details on the formation and analysis associated with the ECDFs and the box plots are discussed below. Analysis of the collected data follows in Section 6.

5.1. Empirical cumulative distribution functions (ECDFs)

The normalized received-power data P_{RX}^{norm} captured at all of the receive sites were combined for each frequency, and a Kaplan-Meier estimate of the ECDF was computed on this aggregate data set with a commercial software package. This ECDF was then used to obtain the parameters for the log-normal CDF. We include some fundamental properties provided in [12] that are relevant to the data analysis.

The log-normal probability density function is given by

$$p(x; \mu, \sigma) = \frac{1}{\sqrt{2\pi}\sigma x} \exp\left[-\frac{(\ln x - \mu)^2}{2\sigma^2}\right], \qquad x > 0, \tag{2}$$

Integrating (2) over $(0, x)$, by use of the substitution $t = -\frac{\ln x - \mu}{\sqrt{2}\sigma}$ and (7.1.2) in [15], yields

$$F(x; \mu, \sigma) = \frac{1}{2}\text{erfc}\left[-\frac{\ln x - \mu}{\sigma\sqrt{2}}\right], \qquad x > 0, \tag{3}$$

where erfc($\cdot$) is the complementary error function, and μ and σ are the mean and standard deviation, respectively, of the natural logarithm of the random variable X. Then, the expected value is found from

$$E[X] = e^{\mu + \frac{1}{2}\sigma^2},\tag{4}$$

the median value is given by

$$\text{median} = e^{\mu},\tag{5}$$

and the standard deviation is found from

$$\text{standard deviation} = e^{\mu + \frac{1}{2}\sigma^2}\sqrt{e^{\sigma^2} - 1}.\tag{6}$$

5.2. Box plots

The "box" in a box plot contains the median value of the data, and the edges of the box correspond to the range of data between two percentiles, typically the 25th (Q_1) and 75th (Q_3) percentiles. A dashed or broken line called a whisker is attached at each end of the box, where the maximum length is typically given by

$$\text{whisker length} = 1.5\,(Q_3 - Q_1),\tag{7}$$

where ($Q_3 - Q_1$) is called the interquartile range, or IQR. If the maximum data value is less than $Q_3 + 1.5(Q_3 - Q_1)$, then the whisker extends only to the maximum data value, which is called the "upper adjacent." Likewise, if the minimum data value is greater than $Q_1 - 1.5(Q_3 - Q_1)$, then the whisker extends only to the minimum data value, called the "lower adjacent." Any data that fall beyond the whiskers are called outliers and are plotted as individual points. The 50th percentile is the median of the data, the 25th percentile is the midpoint of the data below the median, and the 75th percentile is the midpoint of the data above the median. Thus, the symmetry (or asymmetry) of the box provides insight into the symmetrical properties of the data. For example, if the data fit well to a Gaussian distribution, the 50th percentile would equal both the median and the mean, and the box edges and whiskers would be symmetric about that median.

6. Estimated Received Power on Several Floors or Levels of the Structure

Here, we estimate the cumulative distribution of the received signal on several floors or levels within each structure. The point-of-reference is Receive Site 1 for both the subway station and the Empire State Building. These power levels were determined from normalized data with (1), and represent the power received at the external receive site when a 1 W radio transmitter is carried through the building. By normalizing the data to 1 W with a 1 W transmitter as the source, the path loss (*PL*) results are easily found from

$$PL(\text{dB}) = -P_{RX}^{norm}(\text{dBW}).\tag{8}$$

These data provide insight on the expected path-loss and variability that the signal will experience as a function of floor or level.

6.1. Estimating path-loss in the subway

As the transmitter was carried to levels three and four of the subway station, the signal was not discernible at Receive Site 1. That is, the signal-to-noise ratio was less than 0 dB. Thus, to obtain an estimate of the received signal level and corresponding path-loss relative to Receive Site 1, we combined the measured results from Receive Sites 1 and 2. Because Receive Site 2 was located within the subway, we were able to estimate the received mean power on each level in the subway. We then renormalize the data from Receive Site 1 based on knowledge of the path-loss between various levels as determined at Receive Site 2. The steps below explain the process.

1) Normalize the data from Receive Sites 1 and 2 as given in (1).

2) Parse the data into subsets representing the different levels. For the subway, the subsets are
   a. normalized data for Level 1 as measured at Receive Site 1,
   b. normalized data for Level 1 as measured at Receive Site 2,
   c. normalized data for Level 2 as measured at Receive Site 2,
   d. normalized data for Level 3 as measured at Receive Site 2,
   e. normalized data for Level 4 as measured at Receive Site 2.

   The stairwell data were not included in the data for the four levels because the stairwell represents a transition between levels rather than a particular level.

3) Estimate the difference in path-loss between adjacent levels. This is accomplished by computing the absolute difference in the means of the subsets of data for two adjacent levels of data measured at Receive Site 2.

$$PL(\text{dB})^i_{diff} = |\mu^i_{Site\ 2} - \mu^{i-1}_{Site\ 2}|, \qquad i = 2,3,4. \qquad (9)$$

Here, $\mu^i_{Site\ 2}$ is the mean of the subset of data from Receive Site 2 for the ith level.

4) Estimate the mean for Level 1 by use of normalized data measured at Receive Site 1. This is the mean value $\mu^1_{Site\ 1}$ for Level 1, as measured at Receive Site 1.

5) Subtract the difference in path-loss between levels found in Step 3 from the mean path-loss found in Step 4:

$$\hat{\mu}^i_{Site\ 1} = \mu^1_{Site\ 1} - \Sigma^i_2\ PL(\text{dB})^i_{diff}, \qquad i = 2,3,4. \qquad (10)$$

The difference in path-loss between all of the levels that are necessary to reach the desired level must be included, as indicated by the summation in (10). This

45

provides an estimate of the mean for each level if the receiver location was at Receive Site 1. The three values provide an estimate of the mean path-loss with respect to Receive Site 1 for Levels 2, 3, and 4.

6) Renormalize the data for Levels 2, 3, and 4 from Step 2 by subtracting the mean of each parsed, normalized data set. This renormalizes the data for Levels 2, 3, and 4 to a 0 dB mean. The parsed, normalized data from Step 2 has a mean with respect to Receive Site 2, and thus needs to be set to a 0 dB mean before adding the mean path-loss computed in Step 5.

7) Finally, add the path-loss means computed in Step 5 to the renormalized data from Step 6 to obtain data that can be used to estimate the ECDFs and path-loss values.

6.2. Estimating path-loss in the Empire State Building

For the Empire State Building, the normalized data from Receive Site 1 provided sufficient coverage to obtain path-loss estimates for the areas covered in the walk-throughs. In this case, Steps 1 and 2 from the process above were carried out, where "levels" now refers to the "Street," and Floors 1, 5, 20, 29, 61, 80, 83. The normalized data were parsed into subsets corresponding to the different floors of the building, which then were used to create the corresponding ECDF and path-loss estimates.

6.3. ECDF and log-normal results

Based on the process described above, the ECDFs are plotted in Figures 33 and 34 for the various platform levels of the subway station and floors of the Empire State Building. The subplots with the figures correspond to the frequencies used in the measurements, including two frequencies, 750 MHz and 1834 MHz, not utilized by the four RF PASS systems tested. These additional frequencies were included to provide better insight in the frequency dependence of the structures as well as in anticipation of future public-safety usage of those bands, 750 MHz in particular.

Parameters that can be used to generate representative log-normal distributions are found by curve-fitting the data. Table 2 below lists the parameter values for the associated fitted curves in Figures 33 and 34. These values can then be used with equations (2) and (3) to obtain representative PDFs and CDFs for the propagation environments.

Associated with the ECDF plots are box plots shown in Figures 35 and 36. The data are presented as path-loss information in the box plots. These plots provide insight into the symmetry aspects of the data as well at the spread of the data. Table 3 and 4 contain the 25th, 50th, and 75th percentile values for the path-loss results of the subway station and Empire State Building, respectively. Recall that by normalizing the data to 1 W, the path loss and measured power results are simply the inverse of each other.

In Figure 33, the ECDF results fit a log-normal distribution well for Levels 1 and 4 for all five frequencies. ECDF results for the other two levels generally follow a log-normal distribution, but not as well as Levels 1 and 4. This is likely due to the fact that

46

the areas covered on Levels 2 and 3 represented a larger physical distance, with more coverage of the widely separated areas of the platform. The marker numbers and walked-path shown in Figure 7 shows the difference in coverage between the four levels. These coverage differences are also reflected in the spread or range of the received power for each level. For example, at 905 MHz, the Level 1 results range from approximately -140 dBW to -100 dBW, or a 40 dB spread. For Level 2, the range is from -200 dBW to -120 dBW, or an 80 dB spread.

Another interesting feature shown in Figure 33 is that at the frequencies of 905 MHz and lower, the Level 3 and 4 results overlap in the 0.2 to 0.7 probability regions. The 1834 MHz and 2405 MHz results overlap only below the 0.1 probability region. Looking at Figure 35, the impact of the overlap effect becomes more obvious as the box for Level 4, depicting the 25th to 75th percentile coverage, fits within the box boundaries for Level 3 of the three lower frequencies. Figure 35 also shows that the median values decrease for the 1834 MHz and 2405 MHz frequencies as the radio transmitter is carried to the lower levels.

The Empire State Building ECDF results in Figure 34 and associated box plots in Figure 36 show how the signal is attenuated as the radio is carried to ever higher floor levels. In Figure 34, the ECDFs generally follow a log-normal distribution, as is evident by the differences between the measured data and the log-normal fit curves. The largest differences occur for the data collected on the street and are likely due to a combination of a limited number of points and a strong line-of-sight contribution.

Another important observation occurs in the significant attenuation that occurs between the data collected on the street and data within the building on the first floor. For the four frequencies other than 430 MHz, the median path loss is between 20 dB and 30 dB. More interestingly, the median path loss for 430 MHz is nearly 50 dB, even though there is just a 17 dB range in the median attenuation over all of the floors within the building at 430 MHz. This is easily seen in the median path loss values in Table 4. During these measurements, aluminum scaffolding covered the side of the Empire State Building facing Receive Site 1. We hypothesize that the scaffolding caused significant attenuation at 430 MHz, but had much less impact at the higher frequencies. The free-space wavelength at 430 MHz is 0.7 m, and the scaffolding, with all the accompanying supports and safety railings had many features on the same order as this dimension. The performance of the RF PASS appeared not to be impacted because the path-loss was still less than 115 dB for most of the test locations, and the system operating in the 450 MHz band has demonstrated the ability to successfully operate with path-loss values of around 115 dB.

A third observation is that the box plots for the Empire State Building, Figure 36, indicate several outliers for the frequencies of 905 MHz, 1834 MHz, and 2045 MHz. Those path-loss estimates are measured near the noise-floor limit of the measurement system, so they likely represent a limitation of the measurement equipment, and not a true path-loss value. This same behavior, shown in Figure 35, is observed across all frequencies in the Level 1 results for the subway station. However, if the signal cannot be distinguished from the noise, this implies that the signal has experienced a path-loss sufficient to attenuate the signal to at least the noise level, if not greater. (Note that although the noise floor is indicated as a single value, it actually fluctuates. The noise floor limit shown in Figures 12 22 to 28 actually represents a slight margin, approximately 2 dB, above the upper limit of the fluctuation.)

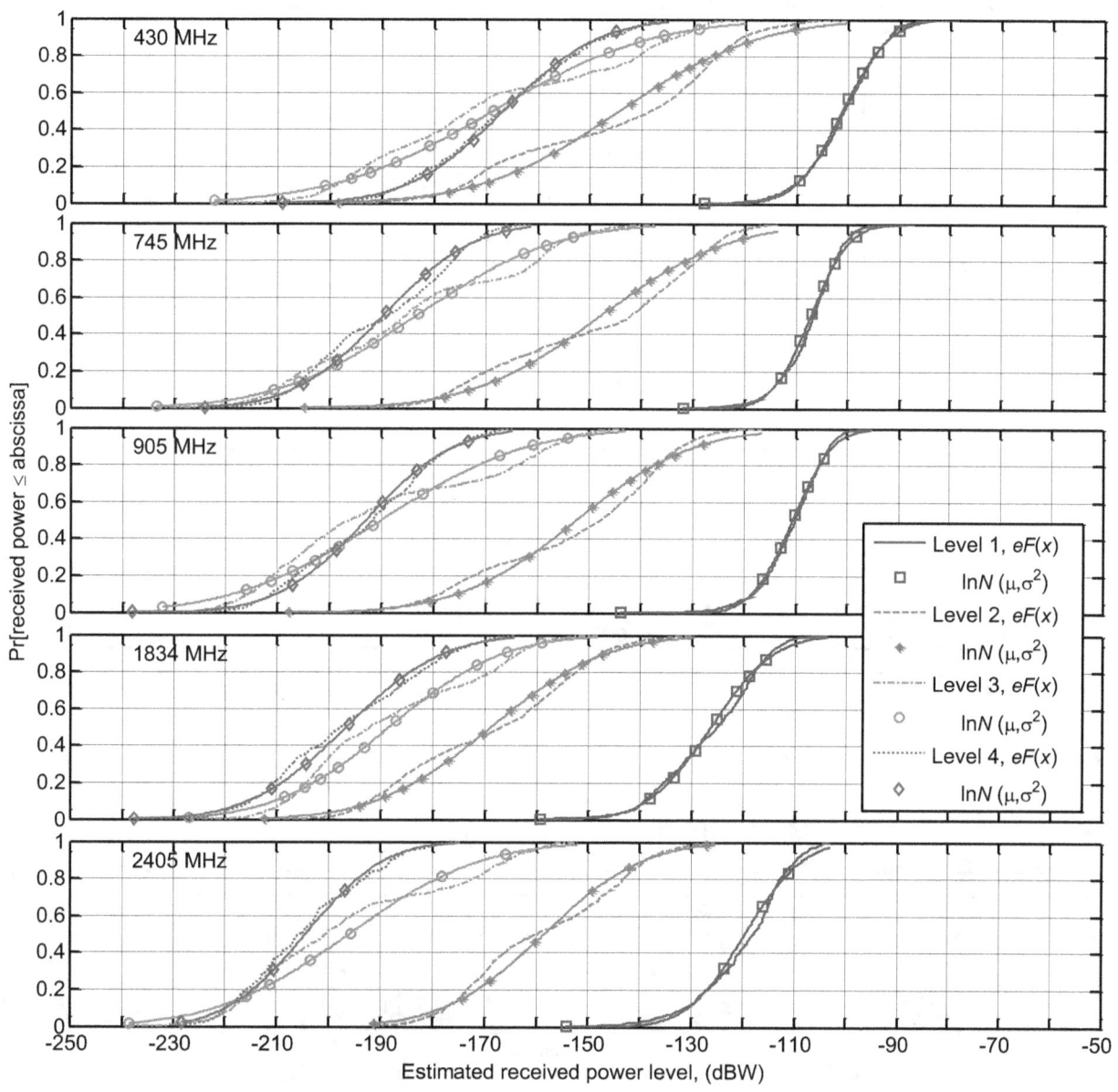

Figure 33. Empirical CDFs and corresponding log-normal-fit curves for the estimated received power across the five frequencies and for the four levels within the subway. Each frequency subplot shows results for the four levels, with the right-most curves representing the first level (and least amount of path loss).

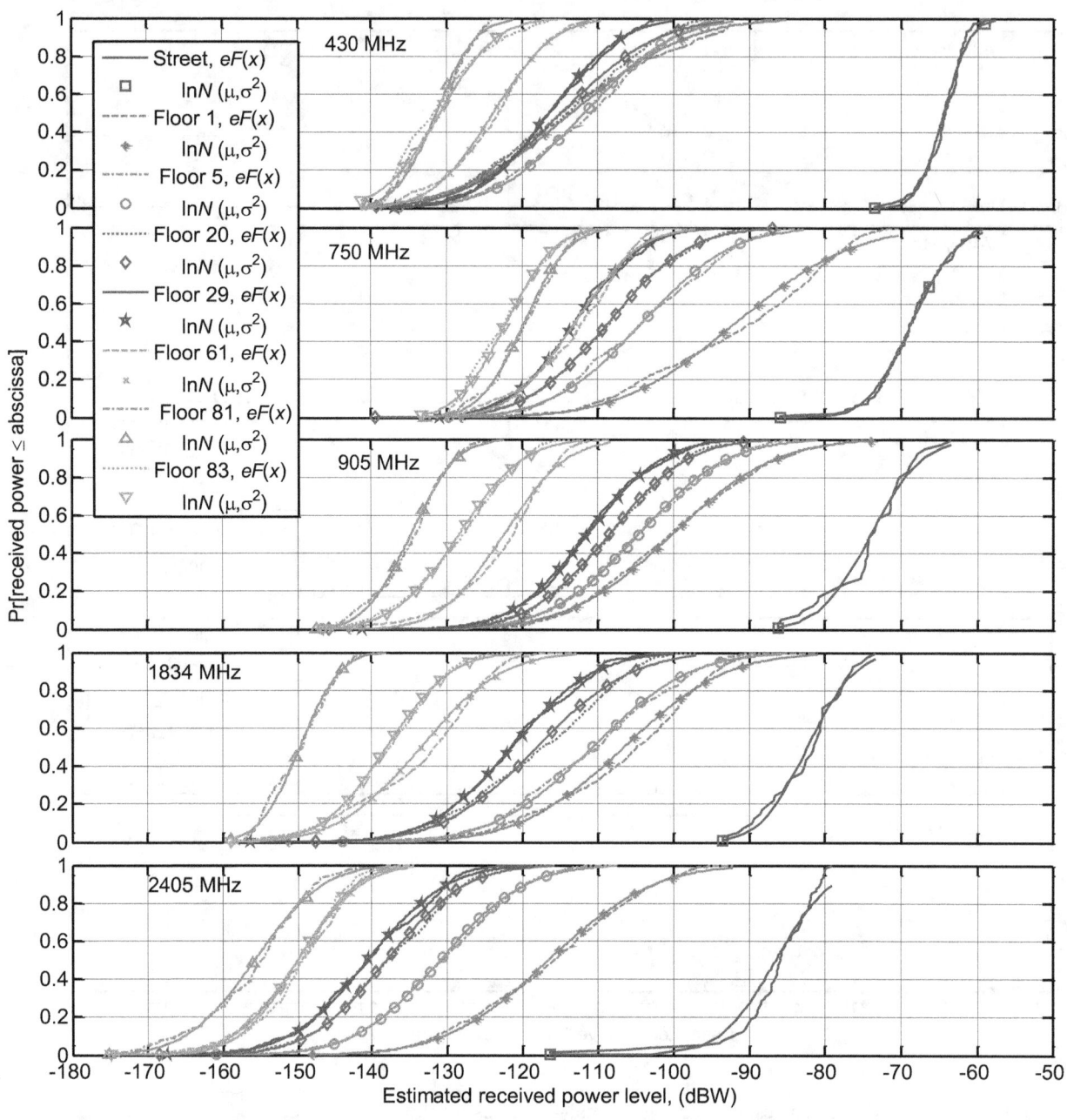

Figure 34. ECDFs representing the estimated received power at the street outside and on the various floors within the Empire State Building.

Table 2. Parameter values to compute the log-normal fit to the empirical CDF data results for the various floors and levels of the subway station and the Empire State Building.

Frequency (MHz)	Parameter Values (dB)							
	μ	σ	μ	σ	μ	σ	μ	σ
	Subway							
	Level 1		Level 2		Level 3		Level 4	
430	-23.3	1.7	-32.1	4.9	-43.4	5.6	-50.7	3.4
745	-24.7	1.4	-32.8	4.5	-45.0	4.8	-51.1	3.1
905	-25.5	1.4	-32.3	4.2	-44.9	5.1	-53.1	3.1
1834	-29.1	2.2	-34.7	4.0	-45.6	3.9	-54.3	3.4
2405	-27.5	2.0	-35.4	3.5	-49.3	4.6	-59.1	2.7
	Empire State Building							
	Street		Floor 1		Floor 5		Floor 20	
430	-14.8	0.6	-26.1	2.8	-25.7	2.1	-26.4	2.3
750	-15.8	1.1	-21.1	2.8	-24.0	2.2	-25.0	2.0
905	-17.0	1.2	-23.1	2.5	-24.1	2.1	-25.0	1.9
1834	-19.0	1.1	-24.5	2.5	-25.5	2.4	-27.2	2.3
2405	-19.9	1.4	-26.8	2.5	-30.1	2.1	-31.8	1.9
	Floor 29		Floor 61		Floor 80		Floor 83	
430	-26.8	1.7	-28.5	1.5	-30.3	1.0	-30.1	1.4
750	-26.0	1.6	-26.0	1.5	-27.6	1.1	-28.1	1.2
905	-25.6	1.8	-28.2	1.5	-31.0	1.1	-29.6	1.6
1834	-28.0	2.0	-30.7	2.0	-34.5	1.0	-31.8	1.6
2405	-32.4	1.9	-34.6	1.6	-35.9	1.7	-34.5	1.4

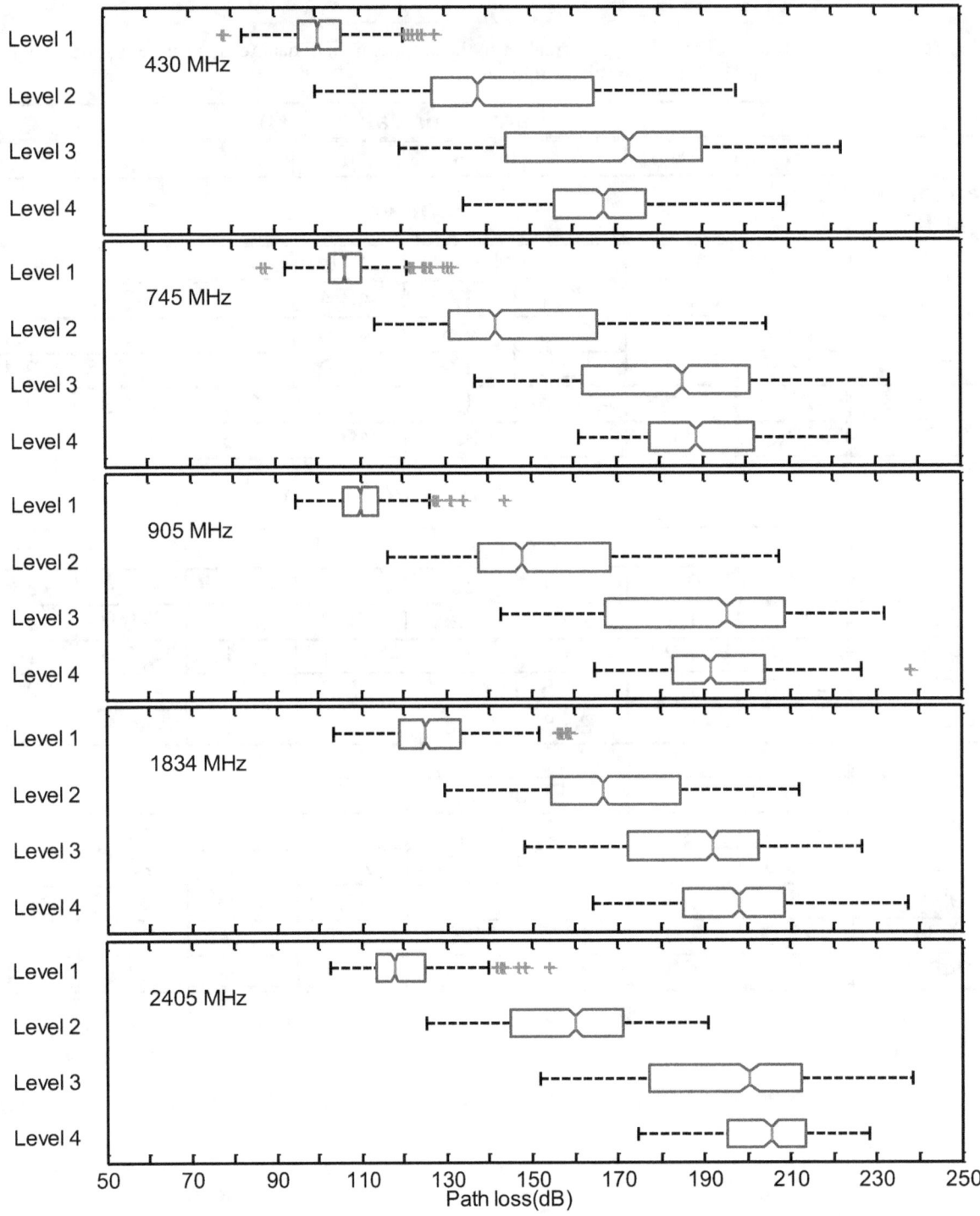

Figure 35. Boxplot of subway path-loss estimates at the four levels. The indent in the box represents the median values, the edges of the box represent the 25th and 75th percentiles, the edges of the whiskers represent the lower and upper adjacent values, and the red crosses are the outliers.

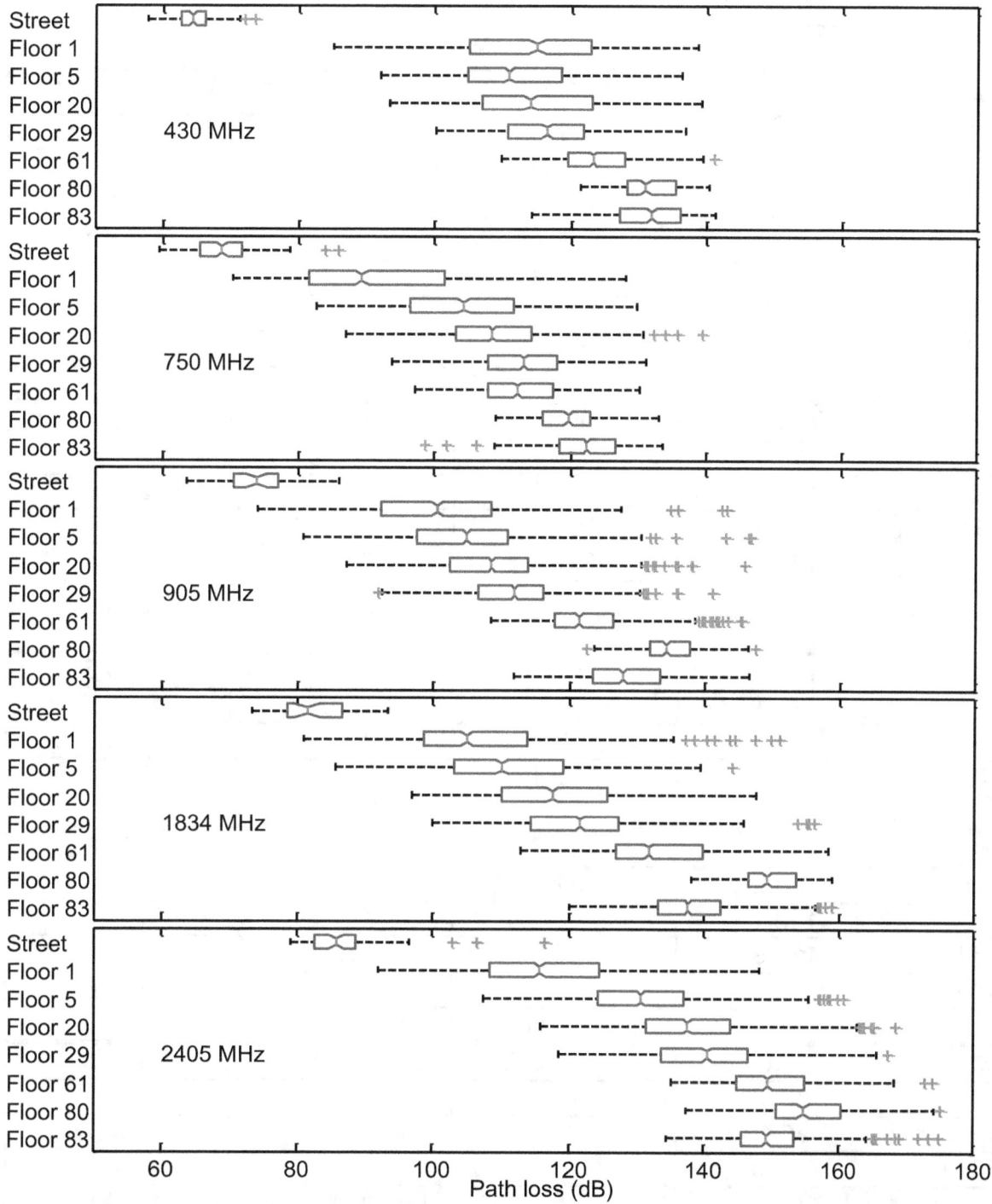

Figure 36. Boxplots of path loss estimates for the Empire State Building. The indent in the box represents the median values, the edges of the box represent the 25th and 75th percentiles, the edges of the whiskers represent the lower and upper adjacent values, and the red crosses are the outliers.

Table 3. Quartiles and adjacents of path-loss estimates for the four levels within the subway. The quartile indicates the percentage of estimated path-loss data below the specified path-loss value.

Frequency (MHz)	Subway			
	Path loss estimates (dB) at quartiles (%)			
	Level 1	Level 2	Level 3	Level 4
	Upper Adjacent			
430	120.6	198.2	222.3	209.0
745	121.2	204.7	233.2	224.0
905	126.4	207.6	232.1	226.6
1834	151.7	212.2	226.8	237.5
2405	139.7	191.0	238.5	228.3
	75 %			
430	106.0	165.2	190.3	177.3
745	110.4	165.7	200.9	201.9
905	114.3	168.5	208.8	204.0
1834	133.4	184.6	202.6	208.6
2405	124.9	171.3	212.4	213.5
	50 % (median)			
430	100.8	138.3	173.4	167.3
745	106.7	142.0	185.3	188.6
905	110.1	147.9	195.7	191.7
1834	125.1	166.7	191.9	198.0
2405	117.8	160.1	200.4	205.4
	25 %			
430	96.1	127.2	144.3	155.8
745	103.2	131.1	162.3	177.9
905	106.1	137.7	167.2	182.9
1834	118.9	154.6	172.4	184.9
2405	113.7	145.1	177.4	195.2
	Lower Adjacent			
430	82.6	100.0	119.8	134.6
745	92.6	113.5	137.2	161.3
905	95.1	116.5	142.8	164.7
1834	103.6	129.5	148.2	164.2
2405	102.8	125.3	152.0	174.8

Table 4. Quartiles of path-loss estimates for the seven floors studied in the Empire State Building. The quartile indicates the percentage of estimated path-loss data below the specified path-loss value.

Frequency (MHz)	Empire State Building							
	Path loss estimates (dB) at quartiles (%)							
	Street	Floor 1	Floor 5	Floor 20	Floor 29	Floor 61	Floor 80	Floor 83
	Upper adjacent							
430	71.2	138.7	136.3	139.3	136.8	139.5	140.5	141.4
750	78.8	127.9	129.7	130.6	131.0	130.0	132.8	133.4
905	86.1	127.5	130.5	130.4	130.3	138.6	146.4	146.6
1834	93.4	135.4	139.4	147.7	145.9	158.3	159.0	156.4
2405	96.5	148.2	155.6	162.9	165.8	168.5	174.4	164.4
	75 %							
430	66.3	122.7	118.5	123.0	121.7	127.9	135.3	136.1
750	71.6	101.5	111.6	114.2	117.9	117.4	122.8	126.5
905	77.1	108.3	110.8	113.7	116.0	126.2	137.8	133.3
1834	86.7	113.7	119.0	125.5	127.2	139.8	153.6	142.5
2405	88.8	124.5	137.1	144.0	146.6	155.0	160.3	153.4
	50 % (median)							
430	64.3	114.9	110.9	113.9	116.4	123.1	130.8	131.8
750	68.7	89.3	104.2	108.4	113.0	112.2	119.5	122.2
905	73.9	100.5	104.7	108.4	111.7	121.3	134.2	127.8
1834	81.5	104.9	110.0	117.4	121.5	131.7	149.4	137.6
2405	85.9	115.7	130.7	137.7	140.5	149.6	154.9	149.4
	25 %							
430	62.6	104.9	104.8	106.9	110.6	119.4	128.1	127.2
750	65.5	81.5	96.4	103.1	107.9	107.9	115.8	118.3
905	70.6	92.2	97.4	102.4	106.5	117.7	131.7	123.4
1834	78.6	98.6	103.1	110.0	114.4	126.8	146.5	133.1
2405	82.6	108.3	124.2	131.5	133.7	144.9	150.9	145.7
	Lower adjacent							
430	57.7	85.1	92.1	93.3	100.1	109.6	121.2	114.2
750	59.4	70.3	82.8	86.9	93.8	97.1	108.9	108.7
905	63.5	74.0	80.9	87.2	92.5	108.4	123.5	111.8
1834	73.3	81.0	85.7	97.0	99.9	112.8	138.1	119.9
2405	79.1	92.1	107.4	115.8	118.5	135.2	137.5	134.4

7. Contributions to Uncertainty in Path Loss Estimates

In this section, we provide an estimate of the uncertainty in our path-loss measurements, as measured by the combination of CW transmitters and the spectrum-analyzer receiver setup. Following the convention described in [16], the uncertainties associated with the measurement and estimation of path loss can be broken into two categories: Type A (evaluated by statistical means), and Type B (evaluated by non-statistical means). Contributions associated with the repeatability of the measurement instrumentation and the transmitted power estimations are described with Type A techniques. Other, systematic effects are described with Type B methods. These include errors in the drift of the measurement instrumentation components not covered in the Type A analysis. We describe these effects below, and then calculate the combined expected uncertainty in our estimation of path loss due to these contributions. This uncertain analysis covers the measurement of a single value, and is, thus, the measurement uncertainty.

7.1 Small-scale fading

One potential source of uncertainty when measuring a propagation channel can be attributed to small-scale fading, often called channel variability in the literature. Small-scale fading occurs due to multiple frequency-, time-, and position-dependent reflections in the local area around each test point. Even though a building environment is fixed and measurements made there would be deterministic, small-scale fading occurs as cars, trucks, and pedestrians move randomly through the environment during measurement. However, the uncertainty analysis here does not consider the channel variability; a relevant discussion on uncertainties attributed to channel variability is found in [17]. Uncertainty estimates due to channel variability in the 434 MHz , 745 MHz and 905 MHz bands are expected to be on the same order of magnitude as the values listed in [17], since the RF propagation environments are similar; i.e., large buildings and urban streets.

7.2 Type A uncertainties of the measurement instrumentation

As discussed in Section 4, CW sources were carried through the structures, and a spectrum analyzer connected to an antenna measured the received power for a single frequency. Use of narrowband filters, data processing, and a check of background environmental signals ensured that the signals we acquired corresponded to the ones we transmitted, rather than those from interfering sources. Measurements from all radios were normalized for nominal differences in transmit power levels (1 W or 5 W, depending on the frequency), as discussed in Section 4.3.

For the measurements here, we performed Type A uncertainty analyses of both the transmitters and the receiver setup. First, using a reverberation chamber, we measured the total radiated power of each radio with the specific transmit antenna attached. The mean results with the relative uncertainty values for the relevant radio/antenna pairings are listed in Table 5. (These values are also displayed in Figure 32)

Table 5. Mean power and relative uncertainty for 10 total radiated measurements of the radio/antenna pairs.

Frequency (MHz)	Antenna	μ (dBm)	u_{TRP}(dB)
430	"rubber duck"	21.5	+1.1,-2.25
745	monopole	26.4	+0.7,-1.0
905	monopole	24.0	+0.6,-0.75
1834	"rubber duck"	24.7	+0.7,-1.1
2400	linear dipole-array	35.3	+0.7,-1.2

The uncertainty analysis of the receiver setup, namely the spectrum analyzer and the laptop used to control the analyzer, consisted of taking six measurements for three different signal level inputs, with a two-hour interval between measurements. The source was a signal generator connected into the spectrum analyzer through a coaxial cable via the port connected to the antenna for the field measurements. The signal generator emitted a CW signal at 0 dBm, and an inline attenuator of 20 dB, 50 dB or 80 dB was included to represent different path-loss conditions. Two minutes of measurement data, consisting of 100 data samples, were collected for each of the frequency bands and attenuator values with the same software and processing as in the field measurements. To approximate the field experiment conditions, the spectrum analyzer and controlling laptop were placed on an external loading dock, which experienced sun in morning and shade in the afternoon. This test was performed during November in Boulder, CO, where the normal temperature range is -1.1° to 13.3° C (30° to 56° F). Because we wished to study the drift of the receiver setup, the signal generator was located inside the building, having more stable environmental conditions.

To obtain the uncertainty value for each frequency, the data were normalized to the mean value of the combined results of the six different tests at each of the three power levels. These normalized data, which now had a nominal 0 dB mean, were then combined, resulting in a data set of 1800 samples. Then, the standard deviation was computed for each frequency set of 1800 samples. Based on the six tests and the three different inline attenuation levels, the relative uncertainty of the receiver setup, $u_{receiver}$, was less than 0.1 dB for all frequencies except 1834 MHz. The 1834 MHz case had a standard deviation of just less than 1 dB due to more variability when the 80 dB attenuator was inline. This was the only combination of frequency and attenuator that demonstrated any significant variability between the six tests intervals.

We also include the uncertainty $u_{analyzer}$, for the spectrum analyzer itself, as provided by the manufacturer and is given as typically less than 0.6 dB.

7.3 Type B uncertainties of the measurement instrumentation

We next consider the Type B uncertainties of the receiver setup for the measurement system. In the section above, statistical methods were used to arrive at uncertainties for various components of the measurement system. The main component not covered in that analysis was the uncertainty contribution due to the RF cables connecting the receive antenna to the spectrum analyzer. Factors that can contribute to uncertainty include items such as temperature drift, the repeatability of connecting cables to the equipment and antennas, and the flexing or bending of the cables. While the impacts due such factors are expected to be small, the uncertainties are discussed because these measurements take place in less than optimally controlled environments. We call this uncertainty u_{drift} and estimate it as less than 0.2 dB, based upon observations over numerous field measurements.

7.4 Combined uncertainty of the measurements

Table 6 summarizes the various components of uncertainty described in the previous subsections. The combined uncertainty in the measured data is then

$$u_{\text{combined}} = \sqrt{u_{\text{TRP}}^2 + u_{\text{receiver}}^2 + u_{\text{analyzer}}^2 + u_{\text{drift}}^2}. \qquad (11)$$

These uncertainties were combined by use of root-sum-of-squares addition on the linear (as opposed to logarithmic) values, and then converted back to decibels, giving u_{combined} of typically less than 1.4 dB, and less than 2.8 dB for the worst case (i.e., the 430 MHz radio.) This combined uncertainty in the measurement system is reasonable for this type of experiment, where the channel variability will be on the order of 7 dB or greater. In addition, the goal of this work is to provide broad classifications of environments whose measured path-loss values range from a few tens of decibels to well over 150 decibels. Thus, a measurement uncertainty of 2.8 dB is acceptable for our application.

Table 6. Description of the measurement uncertainties with associated values.

Type	Uncertainty Description	Method of Estimate	Values (dB)
Type A	Accuracy in spectrum analyzer measurements.	Specified by the manufacturer.	< 0.6 Typical
Type A	Data collection system tests, including laptop and spectrum analyzer.	Collected statistical data for a known source over a one day period, in an outdoor environment.	0.1 (1.0 for 1834 MHz)
Type A	Transmitter reverberation chamber TRP measurements.	Standard deviation of 10 independent calculations of TRP.	0.6 to 2.25
Type B	Cable changes due to temperature.	Observations from previous uncertainty experiments.	< 0.2

8. Summary and Conclusion

RF PASS tests were performed in a New York subway station and the Empire State Building because these types of structures provide challenging RF propagation-channel environments. In the subway, the RF PASS systems were limited in their ability to communicate beyond the initial entrance level. Without the use of repeaters, most of the systems could communicate only a short distance beyond the bottom of the stairwell that connected the token booth corridor to the street. Two systems used repeaters to extend the coverage area. When a repeater was located at the base of the stairwell leading up to the street, those two systems were able to communicate the RF PASS alarms between the street level and the first passenger platform. However, with only a single repeater, neither of the two repeater systems was able to communicate between the external receive site and the second passenger level. This suggests that for structures with sizable subterranean sections, a repeater system will likely be required to reach an external incident command post. If the structure has multiple subterranean levels of increasing depth, a multiple-hop relay system will likely be necessary to ensure the reliability of the communication channel.

In the path-loss measurements and analysis performed at five frequencies, ranging from 430 MHz to 2405 MHz, there are several important insights. Based on the upper adjacent values in the box-plot statistical representation of the path-loss data from the Empire State Building (see Figure 36), path-loss values of 140 dB to 175 dB are possible for high-rises. For the subway, the path-loss values exceed 210 dB to 240 dB at the lower two passenger platforms (see Figure 35). The frequency dependence is more pronounced for the Empire State Building results, but less apparent in the subway data. Thus, while a system may function well at the lower end of the frequency spectrum in the above ground portions of a large building, the subway results demonstrate that subterranean structures can cause path-loss values greater than 200 dB across the 430 to 2400 MHz range.

The testing completed here focused on RF PASS system performance and RF propagation-channel measurements in a high-rise and subway station. While a primary goal of the effort was to look at the correlation between the system performance and path-loss behavior, a secondary goal was to gather path-loss data in two high-attenuation settings. Thus, parameter values for log-normal distributions that will allow simulation of the measured path-loss conditions are included in this report. The authors hope that the data presented here, along with future sets of data, can be used to develop a complete suite of test methods, not only for RF-based PASS systems, but also for other RF-based electronic safety equipment. The path-loss values obtained here are general and could be used to develop standards for other equipment as the need arises for standards for these systems.

Acknowledgements

The analysis and development of test methods for RF-based emergency equipment has been funded by the Department of Homeland Security Standards Branch, Bert Coursey, Chief (retired 2011), Philip Mattson Acting Chief (present).

Several organizations and many people were responsible for the success of this in-field measurement program: The Fire Department of New York provided logistics and support for the field measurements; the New York City Transit Authority facilitated access to the subway station; and Empire State Building personnel provided access for the measurements within the building.

From the New York City Transit Authority, we thank Tom Leonard, Office of System Safety; Paul R. Gerardi, Manager, Fire Safety, Office of System Safety; and Thomas Derienzo, Jr., Superintendent, Fire Suppression, Track and Infrastructure, Maintenance of Way. From the Empire State Building, we thank Karl Tremmel, Chief Engineer; and Timothy J. Clancy, Director of Operations. And from the Fire Department of New York, we owe thanks to Chief Edward Kilduff, Chief of Department; Deputy Assistant Chief Stephen Raynis, Chief of Safety Command; Battalion Chief Robert Keys, Chief in Charge (retired), Research and Development Unit; and the officers and firefighters of FDNY's Research and Development Unit including Lt George Hough, FF George Grammas, FF Mario Tarquinio, and FF Ed Clancy.

Finally, we acknowledge the support received from NIST Physical Measurements Laboratory management, Perry Wilson, RF-Fields Group Leader, and Mike Kelley, Division Chief.

References

[1]   D.R. Novotny, J.R. Guerrieri, and D.G. Kuester, "Potential interference issues between FCC Part 15 compliant UHF ISM emitters and equipment passing standard immunity testing requirements," IEEE EMC Symp. Dig., Aug. 2009, pp. 161-165.

[2]   M.R. Souryal, D.R. Novotny, J.R. Guerrieri, D.G. Kuester, and K.A. Remley, "Impact of RF interference between a passive RFID system and a frequency hopping communications system in the 900 MHz ISM band," IEEE EMC Symp. Dig., July 2010, pp. 495-500.

[3]   K.A. Remley, M.R. Souryal, W.F. Young, D.G. Kuester, D.R. Novotny, and J.R. Guerrieri, "Interference tests for 900 MHz frequency-hopping public-safety wireless devices," IEEE EMC Symp. Dig., Aug. 2011, pp. 497-502.

[4]   National Institute of Standards and Technology, "Statement of Requirements for Urban Search and Rescue Robot Performance Standards-Preliminary Report"http://www.isd.mel.nist.gov/US&R_Robot_Standards/Requirements%20Report%20(prelim).pdf .

[5]   K.A. Remley, G. Koepke, E. Messina, A. Jacoff, and G. Hough, "Standards development for wireless Communications for Urban Search and Rescue robots," Proc. Intl. Symp. Advanced Radio Technologies, June 2007, pp. 66-70.

[6]   National Fire Protection Association, NFPA 1982: Standard on Personal Alert Safety Systems (PASS), document scope available at http://www.nfpa.org/aboutthecodes/AboutTheCodes.asp?DocNum=1982 , accessed Nov. 4, 2011.

[7]   C.L. Holloway, W.F. Young, G. Koepke, K.A. Remley, D. Camell, and Y. Becquet, "Attenuation of Radio Wave Signals Coupled Into Twelve Large Building Structures," NIST Technical Note 1545, Aug. 2008.

[8]   W.F. Young, K.A. Remley, J. Ladbury, C.L. Holloway, C. Grosvenor, G. Koepke, D. Camell, S. Floris, W. Numan, and A. Garuti, "Measurements to support public safety communications: attenuation and variability of 750 MHz radio wave signals in four large building structures," NIST Technical Note 1552, Aug. 2009.

[9]   W.F. Young, C.L. Holloway, G. Koepke, D. Camell, Y. Becquet, and K.A. Remley, "Radio-wave propagation into large building structures—part 1: CW signal attenuation and variability, IEEE Trans. Antennas Propagat., vol. 58, no. 4, Apr. 2010, pp. 1279-1289.

[10]   W. F. Young, K. A. Remley, D. W. Matolak, Q. Zhang, C. L. Holloway, C. Grosvenor, C. Gentile, G. Koepke, and Q. Wu, "Measurements and models for the wireless channel in a ground-based urban setting in two public safety frequency bands," NIST Technical Note 1557, Jan. 2011.

[11]   A. Molisch, "Wireless Communications," Chapter 5, pp. 96-99, John Wiley & Sons Ltd., Copyright 2011.

[12]   E.L. Crow and K. Shimizu, (Ed.), "Lognormal Distributions: Theory and Applications", Marcel Dekker, New York, 1988.

[13]   C. Forbes, M. Evans, N. Hastings, and B. Peacock, "Statistical Distributions," Fourth edition, pp. 131-134, John Wiley & Sons Inc., Copyright 2011.

[14]   J.M. Chambers, W. S. Cleveland, B. Kleiner, and P. A. Tukey, "Graphical Methods for Data Analysis," Chapter 2, pp. 21-24, Duxbury Press, Boston, Massachusetts, Copyright 1983 by Bell Telephone Laboratories Inc.

[15]   M. Abramowitz and I. A. Stegun (Ed.), "Handbook of Mathematical Functions with Formulas, Graphs, and Mathematical Tables," p. 297, Dover Publications, Inc., New York, 1972.

[16]   B. N. Taylor, C. E. Kuyatt, "Guidelines for Evaluating and Expressing the Uncertainty of NIST Measurement Results," NIST Technical Note 1297, September, 1994.

[17]   K.A. Remley, W.F. Young, and J. Healy, "Analysis of Radio-Propagation Environments to Support Standards Development for RF-Based Electronic Safety Equipment," NIST Technical Note 1559, Mar. 2012.

Appendix A: RF PASS Performance and Path Loss Data

Table 7. RF PASS system test results for the subway structure. Associated path loss values are based on the frequency band of operation for that particular system. Note that only Systems 2 and 4 were tested with a single-hop repeater.

PASS test location	System 1			System 2					System 3			System 4				
				Without repeater		With repeater						Without repeater		With repeater		
	Firefighter-down	Evacuation	Path loss (dB)	Firefighter-down	Evacuation	Firefighter-down	Evacuation	Path loss (dB)	Firefighter-down	Evacuation	Path loss (dB)	Firefighter-down	Evacuation	Firefighter-down	Evacuation	Path loss (dB)
1	○	○	67.1	○	○	○	○	66.8	○	○	66.8	○	○	○	○	79.1
2	○	○	94.3	✕	○	○	○	104.7	○	○	104.7	✕	✕	○	○	112.3
3	○	○	104.6	✕	○	○	○	108.1	✕	○	108.1	✕	✕	○	○	120.7
4	○	○	103.9	✕	○	○	○	111.9	✕	✕	111.9	⊗	⊗	○	○	124.7
5	✕	✕	102.5	✕	○	○	○	124.1	⊗	⊗	124.1	⊗	⊗	○	○	132.7
6	⊗	⊗	127.4	⊗	✕	○	○	122.7	⊗	⊗	122.7	⊗	⊗	○	○	144.3
7	⊗	⊗	125.5	⊗	⊗	○	○	138.3	⊗	⊗	138.3	⊗	⊗	✕	✕	153.4
8	⊗	⊗	135.0	⊗	⊗	✕	✕	140.7	⊗	⊗	140.7	⊗	⊗	⊗	⊗	160.7

Legend
- ✕ (square) : Measured failure
- ⊗ (hexagon) : Expected failure
- ○ (square) : Measured success
- ⊙ (hexagon) : Expected success

Table 8. PASS system test results for the Empire State Building. Associated path loss values are based on the frequency band of operation for that particular system.

Legend symbols used in the table:
- ☐○ = Measured success
- ☐✕ = Measured failure
- ⬢○ = Expected success
- ⬢✕ = Expected failure

Test location	System 1 Firefighter-down	System 1 Evacuation	System 1 Path loss (dB)	System 2 Firefighter-down	System 2 Evacuation	System 2 Path loss (dB)	System 3 Firefighter-down	System 3 Evacuation	System 3 Path loss (dB)	System 4 Firefighter-down	System 4 Evacuation	System 4 Path loss (dB)
0	☐○	☐○	58.9	☐○	☐○	61.9	☐○	☐○	61.9	☐○	☐○	69.7
4	☐○	☐○	108.3	☐✕	☐○	91.4	☐○	⬢○	91.4	☐✕	☐✕	100.4
5	☐○	☐○	117.5	☐✕	☐○	99.4	☐○	☐○	99.4	☐✕	☐✕	113.8
8	⬢○	⬢○	105.2	☐✕	☐○	96.6	⬢○	⬢○	96.6	☐○	☐○	119.6
9	⬢○	⬢○	97.7	⬢○	⬢○	93.7	⬢○	⬢○	93.7	☐✕	☐✕	110.3
10	⬢○	⬢○	108.1	⬢✕	⬢✕	106.4	☐○	☐○	106.4	⬢✕	⬢✕	115.8
11	⬢○	⬢○	112.0	⬢✕	⬢✕	106.7	☐○	⬢○	106.7	⬢✕	⬢✕	127.1
12	⬢○	⬢○	103.4	⬢○	⬢○	94.6	⬢○	⬢○	94.6	☐✕	☐✕	114.5
13	☐○	☐○	113.0	⬢✕	⬢✕	107.2	☐✕	☐✕	107.2	⬢✕	⬢✕	123.7
14	☐○	☐○	108.3	☐✕	☐✕	113.2	⬢✕	⬢✕	113.2	☐✕	☐✕	134.2
15	☐○	☐○	109.8	⬢○	⬢○	100.7	⬢○	⬢○	100.7	⬢○	⬢○	121.9
16	☐○	☐○	115.6	⬢✕	⬢✕	108.4	☐✕	⬢✕	108.4	⬢✕	⬢✕	125.9
17	⬢○	⬢○	116.6	⬢✕	⬢✕	106.5	☐○	⬢○	106.5	⬢✕	⬢✕	122.9
18	☐○	☐○	104.3	☐✕	☐✕	103.0	⬢○	⬢○	103.0	☐✕	☐✕	125.8
20	☐○	☐○	112.9	⬢✕	⬢✕	116.2	⬢✕	⬢✕	116.2	⬢✕	⬢✕	126.3
21	☐○	☐○	118.1	⬢✕	⬢✕	106.4	⬢✕	⬢✕	106.4	⬢✕	⬢✕	117.4
23	☐○	☐○	110.9	⬢○	⬢○	103.9	⬢○	⬢○	103.9	⬢✕	⬢✕	135.0
26	☐✕	☐○	128.9	⬢✕	⬢✕	114.1	⬢✕	⬢✕	114.1	⬢✕	⬢✕	137.2
29	☐✕	☐○	136.8	☐✕	☐✕	129.0	⬢✕	⬢✕	129.0	☐✕	☐✕	146.4

Legend	✕ : Measured failure	⬢✕ : Expected failure
	○ : Measured success	⬢○ : Expected success

NIST Technical Publications

Periodical

Journal of Research of the National Institute of Standards and TechnologyCReports NIST research and development in metrology and related fields of physical science, engineering, applied mathematics, statistics, biotechnology, and information technology. Papers cover a broad range of subjects, with major emphasis on measurement methodology and the basic technology underlying standardization. Also included from time to time are survey articles on topics closely related to the Institute's technical and scientific programs. Issued six times a year.

Nonperiodicals

MonographsCMajor contributions to the technical literature on various subjects related to the Institute's scientific and technical activities.

HandbooksCRecommended codes of engineering and industrial practice (including safety codes) developed in cooperation with interested industries, professional organizations, and regulatory bodies.

Special PublicationsCInclude proceedings of conferences sponsored by NIST, NIST annual reports, and other special publications appropriate to this grouping such as wall charts, pocket cards, and bibliographies.

National Standard Reference Data SeriesCProvides quantitative data on the physical and chemical properties of materials, compiled from the world's literature and critically evaluated. Developed under a worldwide program coordinated by NIST under the authority of the National Standard Data Act (Public Law 90-396). NOTE: The Journal of Physical and Chemical Reference Data (JPCRD) is published bimonthly for NIST by the American Institute of Physics (AlP). Subscription orders and renewals are available from AIP, P.O. Box 503284, St. Louis, MO 63150-3284.

Building Science SeriesCDisseminates technical information developed at the Institute on building materials, components, systems, and whole structures. The series presents research results, test methods, and performance criteria related to the structural and environmental functions and the durability and safety characteristics of building elements and systems.

Technical NotesCStudies or reports which are complete in themselves but restrictive in their treatment of a subject. Analogous to monographs but not so comprehensive in scope or definitive in treatment of the subject area. Often serve as a vehicle for final reports of work performed at NIST under the sponsorship of other government agencies.

Voluntary Product StandardsCDeveloped under procedures published by the Department of Commerce in Part 10, Title 15, of the Code of Federal Regulations. The standards establish nationally recognized requirements for products, and provide all concerned interests with a basis for common understanding of the characteristics of the products. NIST administers this program in support of the efforts of private-sector standardizing organizations.

*Order the **following** NIST publicationsCFIPS and NISTIRsCfrom the National Technical Information Service, Springfield, VA 22161.*

Federal Information Processing Standards Publications (FIPS PUB)CPublications in this series collectively constitute the Federal Information Processing Standards Register. The Register serves as the official source of information in the Federal Government regarding standards issued by NIST pursuant to the Federal Property and Administrative Services Act of 1949 as amended, Public Law 89-306 (79 Stat. 1127), and as implemented by Executive Order 11717 (38 FR 12315, dated May 11,1973) and Part 6 of Title 15 CFR (Code of Federal Regulations).

NIST Interagency or Internal Reports (NISTIR)CThe series includes interim or final reports on work performed by NIST for outside sponsors (both government and nongovernment). In general, initial distribution is handled by the sponsor; public distribution is handled by sales through the National Technical Information Service, Springfield, VA 22161, in hard copy, electronic media, or microfiche form. NISTIRs may also report results of NIST projects of transitory or limited interest, including those that will be published subsequently in more comprehensive form.

U.S. Department of Commerce
National Institute of Standards and Technology
325 Broadway
Boulder, CO 80305-3328

Official Business
Penalty for Private Use $300

www.ingramcontent.com/pod-product-compliance
Lightning Source LLC
Chambersburg PA
CBHW081847170526
45167CB00007B/2922